QUANTUM ELECTRONICS AND PARAMAGNETIC RESONANCE

KVANTOVAYA ELEKTRONIKA I PARAMAGNITNYI REZONANS

КВАНТОВАЯ ЭЛЕКТРОНИКА И ПАРАМАГНИТНЫЙ РЕЗОНАНС

The Lebedev Physics Institute Series

Editor: Academician D. V. Skobel'tsyn
Director, P. N. Lebedev Physics Institute, Academy of Sciences of the USSR

Volume 25	Optical Methods of Investigating Solid Bodies
Volume 26	Cosmic Rays
Volume 27	Research in Molecular Spectroscopy
Volume 28	Radio Telescopes
Volume 29	Quantum Field Theory and Hydrodynamics
Volume 30	Physical Optics
Volume 31	Quantum Electronics in Lasers and Masers
Volume 32	Plasma Physics
Volume 33	Studies of Nuclear Reactions
Volume 34	Photomesonic and Photonuclear Processes
Volume 35	Electronic and Vibrational Spectra of Molecules
Volume 36	Photodisintegration of Nuclei in the Giant Resonance Region
Volume 37	Electrical and Optical Properties of Semiconductors
Volume 38	Wideband Cruciform Radio Telescope Research
Volume 39	Optical Studies in Liquids and Solids
Volume 40	Experimental Physics: Methods and Apparatus
Volume 41	The Nucleon Compton Effect at Low and Medium Energies
Volume 42	Electronics in Experimental Physics
Volume 43	Nonlinear Optics
Volume 44	Nuclear Physics and Interaction of Particles with Matter
Volume 45	Programming and Computer Techniques in Experimental Physics
Volume 47	Radio Astronomy: Instruments and Observations
Volume 48	Surface Properties of Semiconductors and Dynamics of Ionic Crystals
Volume 49	Quantum Electronics and Paramagnetic Resonance

In preparation

Volume 46	Cosmic Rays and Interaction of High-Energy Particles
Volume 50	Electroluminescence
Volume 51	Physics of Atomic Collisions

Proceedings (Trudy) of the P. N. Lebedev Physics Institute

Volume 49

QUANTUM ELECTRONICS AND PARAMAGNETIC RESONANCE

Edited by
Academician D. V. Skobel'tsyn
*Director, P. N. Lebedev Physics Institute
Academy of Sciences of the USSR, Moscow*

Translated from Russian

CONSULTANTS BUREAU
NEW YORK–LONDON
1971

The Russian text was published by Nauka Press in Moscow in 1969 for the Academy of Sciences of the USSR as Volume 49 of the Proceedings (Trudy) of the P. N. Lebedev Physics Institute. The present translation is published under an agreement with Mezhdunarodnaya Kniga, the Soviet book export agency.

Library of Congress Catalog Card Number 73-120026
SBN 306-10853-4

© 1971 Consultants Bureau, New York
A Division of Plenum Publishing Corporation
227 West 17th Street, New York, N. Y. 10011

United Kingdom edition published by Consultants Bureau, London
A Division of Plenum Publishing Company, Ltd.
Donington House, 30 Norfolk Street, London, W.C. 2, England

All rights reserved

No part of this publication may be reproduced in any
form without written permission from the publisher

Printed in the United States of America

CONTENTS

MASER INVESTIGATIONS

N. V. Karlov 1

   Introduction 1

   Chapter I. Limiting Sensitivity of Receivers of Electromagnetic Radiation 4
     1. Linear Amplifier and Receiver with Intensity Videodetector 5
     2. The Effect of the Number of Types of Oscillations 8
     3. Spectral Density of Spontaneous Noise for Systems with Negative
        Temperature 9
     4. The Sensitivity of Quantum Counters 11

   Chapter II. Sensitivity of Radio Receivers Using Masers 13
     1. Spectral Density of the Noise at the Input of a Maser in Practical Devices 13
     2. Gain in Sensitivity of a Radio Receiver when Using a Maser 15
     3. Sensitivity of Radio Receivers with a Noncoherent Intensity Detector
        (Square-Law Video Reception) 18
     4. The Technical Threshold of Sensitivity and the System Requirements when
        Using Masers 20

   Chapter III. The Q-Factor of the Active Material 23
     1. Q-Factor and Nondiagonal Elements of the Density Matrix 24
     2. Calculation of the Susceptibility 25
     3. Q-Factor of the Active Material of a Paramagnetic Maser 29

   Chapter IV. Microwave Masers 34
     1. Traveling-Wave Masers 34
     2. The Single-Resonator Maser 37
     3. Two-Resonator Masers 44
     4. Multiresonator Masers 48

   Chapter V. Nonlinear Effects in Masers 52
     1. Saturation and Restoration of the Q-Factor of the Active Material 53
     2. Saturation of Masers 54
     3. Polarization of the Active Material under the Action of the Sum of a
        Strong Monochromatic Signal and a Weak Signal 58
     4. Fields at the Output of a Traveling-Wave Maser 63
     5. The Paramagnetic Maser 66
     6. Analysis of Some Nonlinear Distortions in Traveling-Wave Masers 68

   Chapter VI. Experimental Investigations 72
     1. The Traveling-Wave Maser 72

2. Two-Resonator 21-cm Wavelength Maser Intended for Radio-Astronomy Investigations 77
3. The Use of Rutile in Masers 81
4. Investigation of Calcium—Vanadium Ferrite as a Material for Low-Temperature Nonreciprocal Elements of Masers................................... 85
5. Experimental Investigation of Nonlinear Distortions in a Paramagnetic Maser.. 86

Conclusion 88

References 90

EXPERIMENTS ON ELECTRON PARAMAGNETIC RESONANCE AT TEMPERATURES OF 0.1-4.2°K

V. B. Fedorov... 93

Introduction 93

Chapter I. Electron Paramagnetic Resonance in α,α-Diphenyl-β-picrylhydrazyl Free Radicals at Liquid-Helium and Lower Temperatures 97
1. Review of the Results of Investigations of the Antiferromagnetic Exchange Interaction in Molecular Crystals of Chemically Stable Organic Free Radicals. 97
2. Experimental Results 101
3. Temperature Dependence of the Integral Intensity of the Resonance Absorption Line 102
4. Temperature Dependence of the Resonance Line Width 105
Conclusions 107

Chapter II. Influence of Exchange Interactions on Paramagnetic Relaxation in Magnetically Concentrated Systems 109
1. Experimental Methods for the Investigation of Paramagnetic Relaxation 109
2. Brief Results of the Theory of Paramagnetic Relaxation in the Simplest Systems 111
3. Relaxation through Spin—Spin Degrees of Freedom at Low Temperatures.... 118
4. Investigations of the Relaxation Time of DPPH at Liquid Helium Temperatures at Frequencies of 10^4 MHz and 50 MHz 121

Chapter III. Influence of Exchange Interactions on the Spin-Lattice Relaxation in Magnetically Dilute Systems. Experiments on $K_3(Fe,Co)(CN)_6$ at $T = 0.1-4.2°K$ and 42 MHz Frequency 123
1. Experimental Results 123
2. Two Mechanisms of Spin-Lattice Interaction with Ion Pair Participation 126
3. Hypothesis on the Mechanism of Spin-Lattice Relaxation at Low Temperatures 128
4. Energy Spectrum of Ion Pairs in Potassium Cobalticyanide 129
5. Calculation of the One-Phonon Spin-Lattice Interaction of Exchange-Coupled Iron Ions. Quantative Comparison of the Theory with Experiment......... 133

Chapter IV. Apparatus for the Observation of Electron Paramagnetic Resonance at 42 MHz and below 1°K 136
1. General Layout of the Apparatus 136
2. Special Features of Magnetic Resonance Experiments at Very Low Temperatures 138
3. Construction of the Principal Units of the Apparatus 140

4. Measurement of the Spin-Lattice Relaxation Rate at T < 1°K	140
5. Method for Measuring Temperatures at T < 1°K	142
6. Electric Circuit of the Spectrometer	144
Conclusions	145
References	146

MASER INVESTIGATIONS

N. V. Karlov

Introduction

It is well known that the sensitivity of the various types of radio receivers is limited by the internal noise of the receiver amplifiers. Until the appearance of masers the level of internal noise of receivers in the decimeter, centimeter, and millimeter wave bands considerably exceeded the level of external noise due to the background of cosmic radio radiation, the noise of the atmosphere, and the thermal radiation of the surface of the earth. Masers in the radio wave band have an extremely low level of inherent noise. The effective temperature of this noise usually does not exceed a few degrees Kelvin. It is this low level of inherent noise which is the main property of masers. Hence their practical application leads to a considerable improvement in sensitivity.

An analysis of the possible gain in sensitivity and the conditions for realizing it includes not only an investigation of the improvement in sensitivity when connecting masers to receiver systems. We are not limited to this, since other factors of importance are the broadband properties, the stability of the amplifier, the dependence of the gain on the input signal power, the time for amplification to build up, and the linearity of the amplifier. In addition, the limiting sensitivity of receivers of electromagnetic radiation is also of interest in principle. As a result of the progress on masers in the submillimeter and optical wavebands, an analysis of the limiting sensitivity must be carried out taking into account the possible occurrence in the receivers of many types of oscillation.

First of all it is useful to determine the limiting sensitivity which can in principle be attained when receiving electromagnetic radiation. The development of quantum electronics makes this problem an urgent one. In the classical work by Townes, Shimoda, and Takahashi it was shown that the spectral density of the noise of an ideal maser with one type of oscillation is $h\nu$. Essentially, this is due to the uncertainty relation, which connects the uncertainty in the number of quanta of the radiation with the uncertainty of its phase, and is true for any linear amplifier, i.e., an amplifier in which an increase in the number of quanta of the signal occurs without distortion of the phase relations. In this connection it is of interest to consider the problem of the limiting sensitivity of quantum counters, more accurately, counters of quanta, i.e., receivers in which phase is completely lost but which record the intensity of the radiation information on the signal. In classical electronics a receiver with a video detector at the input is similar to a quantum counter, and a straight amplification receiver or a superheterodyne receiver, being a linear amplifier, is similar to a receiver with a maser.

In this paper we present an analysis of the limiting sensitivity of receivers with linear amplifiers and receivers with video detectors of intensity. Because of the importance of receivers with many types of oscillations this problem is analyzed for the case of N independent types of oscillations. It is shown that the sensitivity of the various receivers is different.

The limiting sensitivity of a detector of intensity corresponds to the reliable recording of one quantum during the time of observation. At the same time for a linear amplifier the least possible level of input fluctuations corresponds to the presence of a background equivalent to one quantum entering the amplifier input in the time of observation. Consequently, a doubling of the output readings indicates, with a probability of the order 60-70%, the appearance of one quantum of useful signal. For reliable recording it is necessary to increase the signal considerably above the level of the output fluctuations.

The dependence of the sensitivity on the number of types of oscillation is important when transferring to the optical band.

The ideal linear amplifier and receiver with a detector of intensity considered above correspond to a quantum amplifier and counter. The active material of masers is a medium the noise of which corresponds to the minimum possible noise of any linear amplifier. At the same time 3-level quantum counters make it possible to record accurately the appearance of a single quantum of signal during the recording time.

Because of the smallness of the inherent noise of masers the use of these amplifiers should lead to a considerable improvement in sensitivity. But we must, however, distinguish between the sensitivity of the masers themselves and the sensitivity of the radio receivers using masers. The first is extremely large. For the second, an important part is played by cosmic noise, atmospheric noise, and the noise in the receiver channels. The part played by these sources of noise is large precisely because the noise of the amplifier is small. Therefore, the problem of connecting masers to receiver systems requires a consideration of the sensitivity of these receivers.

In this paper we obtain relations which enable one to determine the gain in sensitivity which is possible using a maser. When receiving regular signals in the centimeter band one can obtain a gain of 10-50. The expressions obtained enable one to choose the optimal constructive solutions, for which the maximum gain in sensitivity is attained when connecting masers to the antenna systems of practical receivers.

When using masers in radio astronomical equipment, space communication systems, and modern radar complexes, the problem of the phase and amplitude stability of the amplifiers is extremely important due to the high sensitivity of receivers with masers. The problem of the phase stability is particularly important. To solve this it is necessary first of all to analyze the properties of the selectivity of the active material. The formulas for the selectivity of the active material given in the literature do not enable us to carry out this analysis. Therefore, in this paper we make a calculation of the selectivity of the active material of paramagnetic masers with three energy levels and two relaxation times.

By taking into account the nondiagonal element of the density matrix, which corresponds to a free transition, and also by taking into account the detuning of the frequency of the pumping radiation we make this analysis of the stability feasible.

It is found that for a sufficiently deep saturation of the auxiliary transition the phase stability of the Q of the active material is determined in the first place by the stability of the absorption line, i.e., in the case of paramagnetic amplifiers, by the stability of the magnetic field. However, the use of magnets with superconducting windings completely eliminates all instabilities due to possible variations of the magnetic field. As a result, variations of phase can be due only to fluctuations of the parameters of the auxiliary pumping radiation. In this case the variations in the phase of the Q which are due to variations in the pumping frequency are reduced by a factor of T_1/T_2. For an EPR line width $\Delta \nu_l$ equal to 50 MHz, a detuning of the pumping frequency by 5 MHz leads to a change of phase of 10^{-5}-10^{-7} radian. It is obvious that such a high stability of the parameters of the active material of a maser is very important when using masers in high-sensitivity radio receivers.

Although the properties of the active material are important in determining the properties of the maser, many of its parameters depend on the way in which the negative resistance of the active material is realized. For a given gain, depending on the type of amplification system, different passbands are obtained and different amplifier stabilities.

By the successive use of the idea of a complex Q-factor of the material we determine the frequency transmission functions of resonator masers and traveling-wave masers, we analyze the amplitude and phase stability of the amplifiers, and we obtain the equivalent temperatures of the input noise.

It is shown that the noise properties of two- and single-resonator masers, and also traveling-wave masers are the same. Consequently, for application in high-sensitivity receivers multiresonator masers and traveling-wave masers are the most attractive.

The important problem of the bandwidth of masers as it applies to two-resonator masers has been analyzed previously without taking into account the line width of the material. Moreover, at high frequencies the bandwidth is determined not by the Q of the loaded resonator, but by the Q of the line of the material. When constructing wideband systems the width of the line determines the value of the attainable bandwidth in the decimeter waveband also. The effect of the line width becomes important when inversion increases. Therefore, in this article the problem of the bandwidth of masers is analyzed taking into account the line width of the signal transition, and including two- (and more) resonator masers.

When using masers in radio astronomy, besides phase and amplitude instabilities, variations in the shape of the maser frequency characteristics which arise, for example, due to mismatches of the antenna-feeder channels may be important. The parasitic signals which appear in this case may completely nullify the advantages of the high sensitivity. The methods of determining the parameters of masers developed in this article enable us to estimate these parasitic effects and to obtain methods for eliminating them.

In the millimeter and submillimeter wavebands it is difficult to design single-mode high-Q resonators, filled with active material. Here it is more convenient to construct traveling-wave masers. However, in certain cases regeneration of these masers may become feasible by introducing (or incompletely eliminating) reflections from the input and output of the maser. In this article we obtain formulas for the gain and bandwidth of these masers and we analyze their stability.

Masers are linear and are characterized by complex frequency transmission functions. However, this linearity exists only for small levels of the input signals, when the Q of the active material does not depend on the signal intensity. However, because of the effect of saturation when the intensity of the signal increases, the difference in the populations of the signal transition falls, the gain is reduced, and the linearity of the maser is destroyed. The nonlinear properties of masers require special consideration. Strictly speaking, if the linearity is destroyed we cannot use the idea of the Q of an active material. However, in the case of saturation by a monchromatic signal we can obtain the dependence of the population difference on the intensity of the signal and we can assume that this population difference determines the value of the Q at the resonance frequency for a given signal level.

A reduction in gain reduces the sensitivity, and it is necessary to take this into account when considering the questions of the application of masers, particularly in radar stations. No less important is the value of the time taken for amplification to build up after the saturating signal is removed.

We obtain the dependence of the gain on the power of the input signals and we determine the duration of the amplification restoration time.

To determine the nonlinear distortions within the passband of masers we calculate the polarization of a quantum system with three energy levels and two relaxation times under the action of the sum of a weak and a strong monochromatic field. The polarization is determined by means of the density matrix. A knowledge of the polarization enables us to determine the field at the output of the maser. Expressions are obtained for the field at the output of a traveling-wave maser when a strong field at the resonant frequency, accompanied by weak sideband components, is present at the input.

Analysis of these expressions enables us to determine the value of such nonlinear distortions as the intensity of combination-frequency fields, changes in the phase shift of the sideband components, and changes in the depth of modulation when the maser is saturated. The value of these nonlinear effects strongly depend on the detuning frequency.

The nonlinear distortions in masers have an unusual feature. As distinct from the usual radio engineering devices, where a reduction in the gain due to saturation is inevitably connected with a strong nonlinear distortion of the spectrum of the amplified signals, in masers even a considerable saturation does not lead to any serious distortions of the spectrum for frequencies $\Omega \gg 1/T_1$, i.e., over the whole bandwidth. Therefore masers successfully combine high sensitivity, stability, and negligible nonlinear distortions.

The problems of the theory of masers and their applications developed in this article served as the basis for experimental investigations on the development of masers and for the determination of their properties, and also for the practical application of masers in radio astronomy investigations.

We give a description of a series of experimental and theoretical investigations on masers, from which we obtain the limiting sensitivity of receivers of electromagnetic radiation, we clarify the gain in sensitivity when masers are incorporated in receiver systems, and we determine the important properties of masers such as their bandwidth, their stability, the dependence of the gain on the signal power, the gain restoration time after removal of a strong signal, and the value of the nonlinear distortions.

CHAPTER I

Limiting Sensitivity of Receivers of Electromagnetic Radiation

As is well known, the sensitivity of the different types of radio receiver systems is limited by the internal noise of the receiver-amplifier elements of these systems. Until the appearance of masers the level of the internal noise for receivers in the decimeter and millimeter wavebands considerably exceeded the level of the external noise due to the background of cosmic radio radiation, atmospheric noise, and the thermal radiation of the earth. Masers in the radio waveband have an extremely low level of inherent noise, the effective temperature of which does not usually exceed a few degrees Kelvin. It is this low level of inherent noise which is the important property of masers. Hence their practical use considerably improves sensitivity. An analysis of the possible gain in sensitivity and the conditions for realizing this is of great interest. However, before considering the problems of the sensitivity of concrete radio receiver systems with masers, it is best to determine the limiting sensitivity which in principle can be attained in receiving electromagnetic radiation [1-3]. The development of quantum electronics makes this question an urgent one.

§1. Linear Amplifier and Receiver with Intensity Videodetector

When receiving high-frequency electromagnetic oscillations they are converted directly to low-frequency measurable oscillations. It is obvious that the sensitivity of the receiver is largest when it is determined only by the high-frequency noise directly at the receiver input. The level of the high-frequency input noise may be due to thermal radiation of the external background, radiation which is conveyed by the feeders, or the thermal noise inherent in the receiver element. Bearing in mind the applicability of the considerations made here not only in the radio-frequency band, but also for receivers in the submillimeter and optical bands, we will assume that in a high-frequency passband $\Delta\nu$, N independent types of oscillations can be excited in the receivers considered. If the effective temperature of the total thermal noise is equal to T, its spectral density is

$$P_{1\nu} = N\rho(T), \tag{1.1}$$

where

$$\rho(T) \equiv \frac{h\nu}{e^{h\nu/kT} - 1}.$$

At a reduced temperature T down to absolute zero only zero oscillations remain in the radiation field. They play no part in the energy transmission processes and they will not be considered here. Therefore, as $T \to 0$ the level of input noise $P_{1\nu}$ falls to zero and the sensitivity can be very large, but the increase in sensitivity with falling T is not as great for linear amplifiers as for quantum counters.

In those cases when it is not linear amplification and conversion of the radiation frequency which occurs but its recording, i.e., the counting of the number of quanta which make up the radiation, information on the phase of the radiation is completely lost and the sensitivity of the recording device, in principle, may amount to the reliable recording of single quanta in the time of observation. The sensitivity is always worse in linear amplification. In fact, for any process of linear reception, when the amplitude of the received radiation is amplified and its phase is maintained, the uncertainty relation must be satisfied, which connects the uncertainty Δn in the number of quanta of radiation with the uncertainty $\Delta\varphi$ of its phase

$$\Delta n \Delta\varphi \gtrsim 1/2. \tag{1.2}$$

In the final analysis this relation determines the fundamental limit of sensitivity, which always exists for linear amplification [4]. The uncertainty (1.2) is equivalent to the appearance at the input of a linear amplifier with a power gain G of additional noise in one type of oscillation with a minimum spectral density [5] of

$$P'_{2\nu} = \frac{G-1}{G} h\nu. \tag{1.3}$$

In the case of a multimode amplifier this quantity must be increased by a factor of N

$$P_{2\nu} = N \frac{G-1}{G} h\nu. \tag{1.4}$$

Therefore any linear amplifier with a power gain G always possesses noise given by formula (1.4). Besides this there is thermal noise given by expression (1.1). After being amplified G times this noise is fed to the input of a nonlinear element (the second detector). The absence of linear amplification at the high frequency corresponds to G = 1. Therefore, in the general case the reception of electromagnetic oscillations occurs when the spectral density of the noise

$$P_\nu = N[G\rho(T) + (G-1)h\nu] \tag{1.5}$$

is subjected to nonlinear conversion. In the case considered the limiting sensitivity on the assumption of a single source of noise is the spectral density of P_ν. In this case the output fluctuations are determined by the fluctuation in the radiation energy P_ν. In fact, in the case of square-law detection or, what is the same thing, energy measurement with a bolometer, the output readings are proportional to the input power. And if the detector does not add its own noise, the output fluctuations are determined by the fluctuations of the input power. This approach is valid if we take the radiation energy as the quantity which characterizes the phenomenon considered, and if we regard the fluctuations as the deviations of the instantaneous value of the energy from the mean value, determined for equilibrium thermal radiation by Planck's formula. Einstein obtained an expression for the spectral intensity of the fluctuations of the energy of equilibrium radiation. If E_ν is the spectral density of the equilibrium radiation with N types of oscillations, the spectral density of the fluctuations in its intensity equals (see, for example [6])

$$\overline{\delta E_\nu^2} = E_\nu\left(h\nu + \frac{E_\nu}{N}\right). \tag{1.6}$$

In the further analysis we will assume that Einstein's formula (1.6) is applicable to the spectral density of the amplified noise (1.5). Then the root mean-square value of the fluctuations in energy in a relatively narrow and comparatively uniform frequency band $\Delta\nu \ll \nu$ during the averaging time $\tau > 1/\Delta\nu$ is equal to

$$Y \equiv \sqrt{\overline{\delta P_\nu^2} \Delta\nu\tau} = G\sqrt{[\rho(T)+h\nu]\left[\rho(T)+\frac{G-1}{G}h\nu\right]}\sqrt{N\Delta\nu\tau}. \tag{1.7}$$

Knowing the mean-square value of the output fluctuations it is easy to obtain the sensitivity both in relation to the monochromatic signal of power E and in relation to the spectral density of the noise signal E_ν. Due to these signals the increase in the control components of the output current during the time τ equals $GE\tau$ and $GE_\nu\Delta\nu\tau$, respectively. Then in the usual way, taking as the criterion of the distinguishability of the readings of the output device the condition of equality of the increment of the regular components of the mean-square value of the fluctuations, we obtain

$$E = \sqrt{\left[\rho(T)+\frac{G-1}{G}h\nu\right][\rho(T)+h\nu]}\sqrt{N\Delta\nu/\tau},$$
$$E_\nu = \sqrt{\left[\rho(T)+\frac{G-1}{G}h\nu\right][\rho(T)+h\nu]}\sqrt{N/\Delta\nu\tau}. \tag{1.8}$$

In analyzing the sensitivity of a receiver with a linear amplifier we must assume that $G \gg 1$. Then

$$E = [\rho(T)+h\nu]\sqrt{N\Delta\nu/\tau},$$
$$E_\nu = [\rho(T)+h\nu]\sqrt{N/\Delta\nu\tau}. \tag{1.9}$$

Hence we see that even when there is no background of thermal radiation [$\rho(T)=0$] the sensitivity of the device is limited by noise due to the uncertainty relation (1.2). The spectral density of the power of this noise increases as the frequency ν increases. However, it should be borne in mind that in this case the energy of a single quantum of radiation which is equal to $h\nu$ also increases. Therefore, if by the sensitivity we mean the minimum detectable number of signal quanta, which occur in a time τ, when $\rho(T) = 0$ the limiting sensitivity of the receiver with a linear amplifier does not depend on the frequency. By equating $n = GE\tau/h\nu$ to $Y/h\nu$, it is easy to obtain from expression (1.7) that when T equals 0 the minimum detectable number of quanta in a time τ is given by

$$n = \sqrt{N \Delta \nu \tau}. \qquad (1.10)$$

In this relation as always $\Delta \nu \tau > 1$.

If there is no high-frequency gain, G is equal to 1 and from formula (1.8) it follows that the sensitivity of the receiver is given by the relations

$$E = \sqrt{\rho(T)[\rho(T) + h\nu]} \sqrt{N \Delta \nu / \tau}. \qquad (1.11)$$

It may be of interest to detect the change in the energy of a single oscillator of the thermal radiation field $\Delta \rho = E_\nu / N$. The minimum detectable $\Delta \rho$ equals

$$\Delta \rho = \sqrt{\rho(T)[\rho(T) + h\nu]} \sqrt{1/N \Delta \nu \tau}. \qquad (1.12)$$

In these wavebands, where the Rayleigh-Jeans approximation is inaccurate, the spectral density of the thermal radiation is connected nonlinearly with the temperature. Consequently, in these wavebands it is convenient to use effective temperatures for the characteristics of the receiving systems. However, in several cases the temperature sensitivity may nevertheless be of interest. Solving equation (1.12) for the temperature increment δT, due to the change in energy $\Delta \rho$, taking into account expression (1.1) we obtain for $\delta T/T \ll 1$

$$\delta T = \frac{T}{\sqrt{N \Delta \nu \tau}} \frac{Sh^{h\nu/2kT}}{h\nu/2kT}. \qquad (1.13)$$

As might have been expected, as the ratio $h\nu/kT$ increases the temperature sensitivity deteriorates considerably.

When $h\nu \ll kT$ and $N = 1$ the relations obtained reduce to the well-known expressions

$$E = kT \sqrt{\Delta \nu / \tau} \quad \text{and} \quad \delta T = T \sqrt{1/\Delta \nu \tau} \qquad (1.14)$$

for the sensitivity of a radio receiver, the effective temperature of the input noise of which equals T. Note also that, as can be seen from formulas (1.9) and (1.11), when $h\nu \ll kT$ the difference between the cases of the use and nonuse of a linear amplifier before the detector (the square-law detector, bolometer, etc.) disappears. However, when $\rho(T) \to 0$ this difference becomes important. In this case the threshold of sensitivity of the videodetector receiver elements in fact fall to zero which enables us to detect reliably a single quantum in the time of observation.

In fact, writing the formulas for the sensitivity (1.9) and (1.11) in terms of the number of quanta n received in the N types of oscillations in the band $\Delta \nu$ in a time τ, we obtain for the linear and square-law receivers, respectively

$$\begin{aligned} n_l &= (n_b + 1) \sqrt{N \Delta \nu \tau}, \\ n_s &= \sqrt{n_b(n_b + 1)} \sqrt{N \Delta \nu \tau}, \end{aligned} \qquad (1.15)$$

where n_b is the number of quanta in one type of oscillation of this radiation, from the background of which the useful signal is separated. Consequently, when $n_b = 0$ (T = 0) for the recording intensity of the receiving element (the intensity detector) the number of minimum detectable quanta of the signals $n_s = 0$. This means that in the absence of a useful signal there are no readings including fluctuation readings at the output of the detector. The appearance of an output signal is clearly due to the appearance of a signal at the input and not noise. It is precisely in this sense that we can assert that when T = 0 the limiting sensitivity of the intensity detector corresponds to the reliable recording of a single quantum in the time of observation. At the same time for a linear amplifier when $n_b = 0$ (T = 0) and the minimum possible value of $N\Delta\nu\tau = 1$, $n_l = 1$, i.e., the level of the output fluctuations corresponds to the presence of a background equivalent to the arrival at the input of the amplifier of a single quantum in the time τ. Consequently, a doubling of the output reading denotes with a probability of the order of 60-70% (for Gaussian processes) the appearance of one quantum of the useful signal. For accurate recording a large excess of the signal over the level of the output fluctuations is necessary.

§2. The Effect of the Number of Types of Oscillations

We will now consider how the sensitivity of the receivers depends on the number of types of oscillations N. It would appear that for square-law detection, i.e., when the sum of the N independent and uniformly intense random processes are squared, the intensity of the fluctuations at the output of the detector, i.e., the dispersion of the square of the sum, is proportional to N^2. Consequently, the mean-square deviation (1.7) should be proportional to N and not to $\sqrt{N}$. However, in reception, i.e., for nonlinear conversion, the radiation in each of the possible types of oscillation act independently on the nonlinear element. Each characteristic oscillation is independently detected and the results of detection are summed. Therefore, we do not have the square of the sum, but the sum of the squares of N independent and uniformly intense random processes. The dispersion of such a sum is proportional to N. In fact, let the random processes $X_1(t), X_2(t), \ldots, X_N(t)$ be independent and uniformly intense:

$$\overline{X_1^2} = \overline{X_2^2}, \ldots, \overline{X_N^2} = \sigma^2. \tag{1.16}$$

At the output of the detector we have the random process

$$Z(t) = \sum^N X_i^2(t), \tag{1.17}$$

the mean value of which $\overline{Z} = N\sigma^2$, and the dispersion of which is

$$\overline{(Z-\overline{Z})^2} = \overline{\left(\sum_i^N (X_i^2 - \overline{X_1^2})\right)} = 2N\sigma^2. \tag{1.18}$$

This fact explains the results obtained above.

The position is changed for superheterodyne reception when the frequency of the signal is coherently converted to an intermediate frequency. It is obvious that for a sufficiently low intermediate frequency only one type of oscillation can exist at this frequency. The problem arises as to how the spectral density of the input radiation in the case of N types of oscillation at the frequency of the signal is converted to the intermediate frequency. If the same N types of oscillations are excited in the mixer element at the frequency of the coherent heterodyne as are excited at the signal frequency, then irrespective of the dependence on the heterodyning mechanism the independent characteristic oscillations are independently transferred by the heterodyne into one type of oscillation at the intermediate frequency. In this type of oscillation the intensities of all the N types of oscillations are combined. Consequently, at the input of the intermediate-frequency amplifier there is fed a spectral density

$$S_\nu = \gamma N [G\rho(T) + (G-1)h\nu], \tag{1.19}$$

where γ is the conversion slope, assumed the same for all the N types. Suppose the inherent noise of the intermediate-frequency amplifier and the mixer can be neglected. Then applying Einstein's formula to the spectral density of the noise at the amplifier output (in one type of oscillation)

$$P_{\nu_i} = G_i S_\nu + (G_i) h\nu_i, \tag{1.20}$$

where the subscript i relates to the intermediate frequency and $G_i \gg 1$, we obtain the root mean square value referred to the input of the intermediate-frequency amplifier

$$Y_i = (S_\nu + h\nu_i)\sqrt{\Delta\nu\tau}. \tag{1.21}$$

Formulas (1.19) and (1.21) enable us to obtain expressions for the sensitivity

$$\begin{aligned} E &= N\left(\rho + \frac{G-1}{G}h\nu + \frac{h\nu_i}{\gamma GN}\right)\sqrt{\Delta\nu/\tau}, \\ E_\nu &= N\left(\rho + \frac{G-1}{G}h\nu + \frac{h\nu_i}{\gamma GN}\right)\sqrt{1/\Delta\nu\tau}. \end{aligned} \tag{1.22}$$

These relations were published in [2] for N = 1, also neglecting the term $h\nu_i/\gamma GN$. The relations obtained differ considerably from the corresponding formulas of the previous section by a different dependence on the number of types of oscillations N, which is due to detection in one type of oscillation. If in the superheterodyne reception one uses not the difference but the sum "intermediate" frequency, detection occurs in the same number of types of oscillations as are in the heterodyne and the formulas of the preceding section remain true.

The proportionality of the mean-square deviation of the readings of the output device to $\sqrt{N}$ confirms the independence of the detection of the different types of oscillation; in the same way for the reception of noisy signals the proportionality of this deviation to $\sqrt{\Delta\nu}$ leads to an increase in sensitivity as N increases. Note that the improvement in sensitivity by a factor of $\sqrt{2}$ when receiving two independent types of oscillation, which arises from formulas (1.12) and (1.13), is realized in radio astronomy by receiving two orthogonal polarizations.

In these cases an increase in the number of types of oscillations is equivalent to an increase in the passband of the high-frequency radiometer. Superheterodyne reception is deprived of this advantage. On the other hand, when receiving monochromatic signals, in the case of superheterodyne reception it is easier to reduce the bandwidth of the linear part of the receiver down to a value determined by the conditions of information transmission in each concrete case. Moreover, if the oscillations of the coherent heterodyne are excited in the mixer but not in all the N types of oscillations, i.e., $N_2 < N$, then the spectral density of the N_2 types of oscillations at the high frequency are transferred into one type of oscillation at the intermediate frequency. Therefore the level of noise S_ν is reduced. If in this case the antenna-feeder channel of the receiver is so coupled to free space that the whole of the monochromatic signal which falls on the antenna aperture is propagated through the channel in one type of oscillation, the excitation of a small number of types of oscillations in the mixer by the coherent heterodyne cuts out the noise of those types of oscillations in which there is no signal and, in proportion to N/N_2, considerably improves the sensitivity of the device.

§3. Spectral Density of Spontaneous Noise for Systems with Negative Temperature

In the case of masers the mechanism by which the additional noise (1.3), which is characteristic of linear amplifiers, occurs is spontaneous radiation. When analyzing the spontaneous noise of masers in [7-9] the authors first calculated the probability of spontaneous radiation, and then its spectral density. It is more convenient to consider spontaneous noise from the general point of view, introducing the idea of the effective temperature for stationary nonuniform media [10].

It is well known that the noise properties of physical systems which are in thermodynamic equilibrium are described by the fluctuation-dissipation theorem of Callen and Welton. This theorem connects the spectral density of nonequilibrium fluctuations in a system with its absolute temperature and the imaginary part of the admittance of the system, which characterizes the losses in the system. For nonequilibrium systems the effective temperature is introduced so that it describes the fluctuation in the parameter of the system which is of interest. Masers, which belong to the class of essentially nonequilibrium systems, are characterized by a negative effective temperature, which is a measure of the inversion attained. By definition this temperature equals

$$T_{eff} = \frac{h\nu}{k \ln n_i/n_j}, \qquad (1.23)$$

where n_i is the number of particles in the lower energy level, and n_j is the number of particles in the upper energy level. When there is population inversion $n_i < n_j$ and $T_{eff} < 0$. In writing formula (1.23) we have assumed that the degree of degeneracy of the i and j levels are the same.

The introduction of the idea of T_{eff}, which is an additional quantity for describing inversion, is justified from the point of view of describing fluctuations in the system considered. The temperature T_{eff} connects only those two energy levels between which transitions occur at the signal frequency.

Generalization of the Callen–Welton theorem to the case of stationary nonequilibrium systems, which possess a negative effective temperature and negative energy losses, leads to the statement that the spectral intensity of fluctuations in a system is determined by the values (moduli) of the negative temperature and the negative losses [10]. For systems with one type of oscillation the spectral density of the square of the fluctuation voltage $\bar{\varepsilon}_\nu^2$, generated by a resistance R, equals

$$\bar{\varepsilon}_\nu^2 = 4R \frac{h\nu}{e^{h\nu/kT} - 1}. \qquad (1.24)$$

Since $R = -|R_B|$ and $T = -|T_{eff}|$, we have

$$\bar{\varepsilon}_\nu^2 = 4|R_B| \frac{h\nu e^{h\nu/k|T_{eff}|}}{e^{h\nu/k|T_{eff}|} - 1}, \qquad (1.25)$$

where R_B is the negative resistance corresponding to the negative losses of the active material of the maser. Relation (1.25) is a generalization of the well-known Nyquist formula for systems of this kind. The spectral density of the fluctuation voltage (1.25) corresponds to the spectral density of the noise power

$$P_{\nu sp} = h\nu \frac{e^{h\nu/k|T_{eff}|}}{e^{h\nu/k|T_{eff}|} - 1}. \qquad (1.26)$$

When $h\nu \ll k|T_{eff}|$ this expression reduces to the simple relation

$$P_{\nu sp} = k|T_{eff}|, \qquad (1.27)$$

which is exactly analogous to the formula for equilibrium noise when $h\nu \ll kT$.

In the case of N types of oscillations the spectral density (1.26) must be multiplied by a factor N. In addition, the expression for $P_{\nu\,sp}$ can be given a simpler form by substituting in formula (1.26) the expression (1.23) for T_{eff}. For N types of oscillations

$$P_{\nu\,sp} = Nh\nu \frac{n_j}{n_j - n_i}. \tag{1.28}$$

As might have been expected, the minimum values of the spectral density of spontaneous noise (when $n_j \gg n_i$) and the noise of a linear amplifier (when $G \gg 1$) are the same [see formula (1.4)].

Therefore, it follows from the considerations in this section that the active material of a maser is a medium the noise of which in principle corresponds to the minimum noise of any coherent amplifier.

§4. The Sensitivity of Quantum Counters

We will now consider quantum counters. In these devices there is no linear amplification of the signal, and there is no population inversion and no spontaneous radiation noise. In quantum counters the quanta of the received radiation are converted into very high frequency quanta which correspond to visible or ultraviolet light. These quanta can be recorded directly by means of suitable photoelectric cells. The principle of operation of quantum counters is analyzed in detail in [11]. Here it is convenient to consider the sensitivity of quantum counters as it applies to the general problem of the limiting sensitivity of receivers of electromagnetic radiation.

We will consider three-level quantum counters, the operating material of which is at a temperature which is sufficiently low so that in the absence of an external signal all the particles are in the lower level (level 1) and there is no noise due to spontaneous radiation. When a signal acts on the system the particles transfer to an excited level (level 2), from which specially applied additional radiation transfers them to the higher level 3. The particles transfer spontaneously from this level to the lower state with the emission of a high-frequency light quantum. The high-energy high-frequency quanta are recorded by a photomultiplier capable of recording separate quanta. Obviously in such a process of conversion of the quanta energy the phase of the received radiation is lost and the quantum counter is essentially similar to a bolometer, a square-law detector, or any other noncoherent indicator of intensity.

The number of particles in the third level of the three-level counter in the case of saturation of the auxiliary transition $2 \to 3$ equals [11]

$$n_3 = (W_{12}\tau_{21} + e^{-h\nu_{21}/kT}) n_0, \tag{1.29}$$

where W_{12} is the probability of a transition induced by the signal of frequency ν_{21}, τ_{21} is the lifetime of particles in the second level, T is the temperature of the operating material of the counter, and n_0 is the total number of received particles. It is well known that the probability W_{12} is proportional to the number of quanta of electromagnetic radiation at the signal frequency. In the absence of radiation at the frequency of the transition $2 \to 1$ when $T = 0$, $n_3 = 0$. At a fairly low temperature the value of n_3 is determined by the background of radiation at the frequency ν_{21}.

Due to spontaneous radiation particles transfer from level 3 to level 1 emitting quanta $h\nu_{31}$. The photocell detects these spontaneous transitions and emits photoelectrons. The photoelectrons form shot noise of random video pulses in the photocurrent. Each pulse in the current is due to one photoelectron. This shot noise is complicated by the fact that the light which falls on the photodetector is spontaneous radiation which has a noisy form. The part played by the fluctuations in the instantaneous intensity of the light P(t) can be taken into account if, following [12], we assume that the probability of the emission of one electron in a time dt $P_1(t_1 dt)$ equals

$$P_1(t_1 dt) = \alpha P(t) dt, \tag{1.30}$$

where α is the quantum sensitivity of the photoelement. This relation reduces to the probability distribution which is a generalization of the well-known Poisson formula.

We see from expression (1.30) that the average value of the number of photoelectrons emitted in unit time is

$$\bar{n}_1 = \alpha \overline{P(t)} = \alpha h \nu_{31} \frac{n_3}{\tau_{31}}. \tag{1.31}$$

The spectral density of the intensity of the fluctuations of noise current, determined by the probability of the single event (1.30), is calculated in [13]. The accurate form of this spectrum is determined by the shape of the line of spontaneous radiation and the photocurrent pulses. This spectrum is uniform for the recording frequencies $\nu < \Delta\nu_{sp} = 1/\pi \tau_{31}$ of interest to us. Its density can be determined by taking into account that the intensity of the spontaneous radiation $P(t) = E^2(t)$, and by assuming that the intensity of the electric field of the light wave $E(t)$ is a random process with a normal probability distribution law. The correlation time of this process equals τ_{31}. For processes of this type the density of the uniform part of the spectrum is given by

$$S(0) = \bar{n}_1(1 + \bar{n}_1 \tau_{31}). \tag{1.32}$$

Taking expression (1.31) into account and assuming that for high-power quanta $h\nu_{31}$ the product $\alpha h \nu_{31} = 1$, we obtain

$$\bar{n}_1 = n_3/\tau_{31}, \qquad S(0) = (1 + n_3) n_3/\tau_{31}. \tag{1.33}$$

For high-power quanta of spontaneous radiation the sensitivity of photoelements (photomultipliers) is so large that the video-amplifier noise can be neglected. Relations (1.33) then enable us to obtain the sensitivity of a device which measures a photocurrent with a time constant $\tau > \tau_{31}$, to a variation in the number of particles in the third level of the counter

$$\delta n_3 = \sqrt{n_3(1+n_3)} \sqrt{\tau_{31}/\tau}. \tag{1.34}$$

Therefore [see the discussion on formula (1.15)], formula (1.34) proves the possibility of the reliable recording of the appearance of a single particle in the third level of the counter. In turn, at a sufficiently low temperature n_3 is proportional to the intensity of the electromagnetic radiation, which is incident on the input of the quantum counter in N types of oscillations and in a frequency band corresponding to the width of the absorption line $\Delta\nu_l$ of the transition $1 \to 2$. Consequently, by measuring n_3 it is possible in principle to record reliably the appearance of a single quantum of the signal in the recording time τ.

The statements which we made earlier [14] about the circuit of a wide-band receiver with a nonlinear quantum converter applies to the quantum counter. In this a change in the population of the upper level of the signal transition is recorded as a change in the intensity of the absorption line of the auxiliary transition. In this case the signal and auxiliary transition have a common upper level, and the frequency of the signal exceeds the frequency of the auxiliary transition. The sensitivity of such a device can in principle also be brought up to the level where it can record a single quantum in the time of observation.

Hence quantum electronic devices — masers and quantum counters — possess a sensitivity which is obtainable in the limit for receivers with linear amplifiers and with intensity detectors respectively.

CHAPTER II

Sensitivity of Radio Receivers Using Masers

Because of the smallness of the inherent noise of masers their use can considerably improve the sensitivity of radio receivers. However, we must differentiate between the sensitivity of the maser itself and radio receivers using masers. The first is extremely large. In the second the noise from space, of the atmosphere, from the surface of the earth, and of the antenna-feeder channels plays a large part in determining the overall level of sensitivity. The part played by this noise is large precisely because the inherent noise is small. In this chapter we derive relations which enable us to estimate the gain in sensitivity which is possible when using masers in receiver systems intended for different purposes [1, 15, 16], and we consider the possibilities which are opened up by the use of broad-band (high-frequency) and relatively high-sensitivity noncoherent detectors of intensity.

§1. Spectral Density of the Noise at the Input of a Maser in Practical Devices

Bearing in mind, as in section 1, chapter I, the applicability of the latter considerations not only in the radio band but also in the submillimeter and optical band, we will assume that in the receiver channels of the devices considered N independent types of oscillations can be excited. In addition, we will assume that the antenna-feeder channel is matched to the input of the amplifier for all the N types of oscillations.

First of all, in addition to the inherent noise of the maser, thermal radiation of the antenna-feeder channel which is at a temperature T_t enters the input. If there are N types of oscillations in the channel each of which propagate with the same attenuation α, then in accordance with the waveguide form of Kirchhoff's law [17] the spectral density of the thermal radiation of the channel at the output equals

$$P_{\nu t} = N\alpha \frac{h\nu}{e^{h\nu/kT_t} - 1} \,. \tag{2.1}$$

In the radio band for N = 1 and $h\nu \ll kT$ the effective temperature of the noise in the channel equals

$$T_{t,\text{eff}} = \alpha T_t \,. \tag{2.2}$$

If the channel is at room temperature, for losses in the channel of 3 dB the effective temperature of its inherent thermal radiation is approximately 150°K.

In addition, the antenna receives thermal radiation of the external background from which the received signal is separated. We will assume that the brightness temperature of the background T_b is fairly constant within the limits of the directional pattern of the antenna system. The antenna can be replaced by its radiation resistance placed at the input of the antenna-feeder channel. Then it is clear from thermodynamic considerations that this resistance is heated by the background radiation to a temperature T_b. The equilibrium radiation corresponding to T_b is re-emitted by the resistance into the channel. For N types of oscillations the total spectral density equals

$$P'_{\nu b} = N \frac{h\nu}{e^{h\nu/kT_b} - 1} \,. \tag{2.3}$$

We can come to the same conclusion by considering the flow of energy of the radiation of a black body in an isotropic medium

$$S = \frac{cU}{4}, \tag{2.4}$$

where c is the velocity of light, and U is the volume density of the spectral density of the radiation. Since

$$U = 4\frac{v^2}{c^3}\frac{h\nu}{e^{h\nu/kT_b}-1}, \tag{2.5}$$

we have

$$S = \frac{1}{\lambda^2}\frac{h\nu}{e^{h\nu/kT_b}-1}. \tag{2.6}$$

In one type of oscillation an antenna with effective receiving area $A = A(\theta, \varphi)$, where θ and φ are the polar angles, receives a power

$$P_\nu = \int_\Omega SA(\theta, \varphi)\,d\Omega = S\int_\Omega A(\theta, \varphi)\,d\Omega, \tag{2.7}$$

where integration is carried out over the whole solid angle Ω. We know from antenna theory (see, for example, [18]), that for any diffraction directional pattern

$$\int_\Omega A(\theta, \varphi)\,d\Omega = \lambda^2. \tag{2.8}$$

Consequently, the spectral power $h\nu/(e^{h\nu/kT_b}-1)$ is received in one type of oscillation. If the antenna is matched to space for N types of oscillations we obtain formula (2.3).

In the case of receiving systems with large aperture and with channels of large cross section, when the density of the oscillators of the field in the receiver practically coincides with the density of the types of oscillations in free space, the received power equals the product of the flow of energy and the geometrical area of the receiving aperture A

$$P'_{\nu b} = \frac{A}{\lambda^2}\frac{h\nu}{e^{h\nu/kT_b}-1}. \tag{2.9}$$

The radiation applicable to the antenna is attenuated by a factor of $(1-\alpha)$ due to loss in the channel. Therefore, the spectral density of the background radiation at the output of the channel equals

$$P_{\nu b} = N(1-\alpha)\frac{h\nu}{e^{h\nu/kT_b}-1}. \tag{2.10}$$

When the channel is matched to the input of the maser for all N types of oscillations the spectral density (2.10) and (2.1) are added to the inherent noise of the amplifier. Therefore, the spectral density of the noise and the input of the amplifier is

$$P_\nu = N\left((1-\alpha)\frac{h\nu}{e^{h\nu/kT_b}-1} + \alpha\frac{h\nu}{e^{h\nu/kT_t}-1} + \rho_{\nu a}\right), \tag{2.11}$$

where $\rho_{\nu a}$ is the spectral density of the inherent noise at the input of the amplifier in one type of oscillation.

In the radio band and for N = 1 the effective noise temperature at the input of the amplifier equals

$$T = (1-\alpha)T_b + \alpha T_t + T_a, \tag{2.12}$$

where the effective noise temperature of the amplifier T_a is due both to the inherent noise of the amplifier (see § 3, Chapter III), and to the noise of the radio receiver used together with the maser. Under practical conditions the latter noise, generally speaking, cannot be neglected. If the noise factor of the receiver is F, the effective temperature of its noise is (F − 1), where T_s = 300°K. Referred to the input of the amplifier with a power gain G we obtain the value (F − 1) T_s/G, which must be added to the noise temperature at the amplifier input (1.27). [Later (see Chapter IV) we will calculate the noise temperature of cavity masers and traveling-wave masers and we will determine the conditions of applicability of formula (1.27)]. Then T_a = T_{eff} + (F − 1)T_s/G, and referred to the input of the antenna-feeder system the effective noise temperature of the radio receiver with a maser is

$$T_{in} = T_b + \frac{\alpha}{1-\alpha} T_t + \frac{F-1}{G(1-\alpha)} T_s + \frac{1}{1-\alpha} |T_{eff}|. \tag{2.13}$$

It is this temperature which determines the sensitivity of the receiver.

The quantity $|T_{eff}|$ does not exceed a few degrees Kelvin, the losses α in principle can be very small, and the gain G can be very large. The sensitivity will then be determined by the value of T_b. This value is due to the thermal radiation of the earth's surface, received by the back and side lobes of the directional pattern, the thermal radiation of the atmosphere, and the background of galactic radio radiation. Data on the dependence of the radiation of the atmosphere and the galaxy on frequency are well known. An analysis of these leads to the conclusion that the best sensitivity can be obtained in the frequency band 1-10 GHz. At very long wavelengths the background of galactic radio radiation is large, and at very short wavelengths atmospheric absorption and, consequently, atmospheric radiation is large. However, it should be borne in mind that at heights exceeding several kilometers the thermal radiation of the atmosphere is close to zero even in the submillimeter band.

§2. Gain in Sensitivity of a Radio Receiver when Using a Maser

When considering the sensitivity of radio receivers with masers it is necessary to distinguish between receivers intended for the reception of regular (noiseless) signals and receivers intended for noisy signals.

Knowing the value of T_{in}, it is easy to determine the sensitivity of a radio receiver with a maser when receiving regular signals. In the radio band the minimum detectable intensity is

$$E = kT_{in} \sqrt{\Delta\nu/\tau}, \tag{2.14}$$

where T_{in} is given by formula (2.13). It is obvious that the sensitivity is greatest under those conditions when T_{in} is a minimum.

An estimate of the gain in sensitivity which is possible by using masers is of interest. It is easy to determine the gain when receiving regular signals. In this case the passband of the receivers compared must be the same, since the choice of the band of received frequencies is determined by the spectral composition of the signal, the frequency stability of the transmitter and local oscillator, etc. Hence the gain in sensitivity is equal to the ratio of the effective temperatures of the receivers compared referred to the antenna. Consequently, the ratio of the intensity of the minimum detectable signal in the absence of a maser E_0 to the intensity E of the signal received using a maser equals

$$q = \frac{T_b/T_s + \dfrac{\alpha_0 T_t/T_s + F - 1}{1 - \alpha_0}}{T_b/T_s + \dfrac{\alpha T_t/T_s + (F-1)/G + |T_{eff}|/T_s}{1 - \alpha}}. \tag{2.15}$$

Here α_0 is the absorption coefficient of the antenna-feeder channel without a maser, and $\alpha = \alpha_0 + \alpha_1 - \alpha_0\alpha_1$ is the absorption coefficient in the channel when there is a maser present. The difference between α_1 and α_0 is due to losses in the auxiliary supply lines, and to the ferrite circulator, if one is used, etc. Under practical conditions $F-1 \gg \alpha_0 T_t/T_s$, and if special steps are not taken to cool the channels then $\alpha T_t/T_s \gg |T_{eff}|/T_s$. In addition, under these conditions $T_t = T_s$, and we obtain

$$q = \frac{T_b/T_s + (F-1)/(1-\alpha_0)}{T_b/T_s + \frac{\alpha + (F-1)/G}{1-\alpha}}. \qquad (2.16)$$

Under typical conditions in the centimeter band when making observations at small angles to the horizontal, relation (2.16) leads to a 10-fold gain in sensitivity. For large angles the value of the gain reaches 50-80.

In the decimeter band the residual effect of noise of the radio receiver used together with the maser is, as a rule, negligibly small. Then

$$q = \frac{T_b/T_s + (F-1)/(1-\alpha_0)}{T_b/T_t + \alpha/(1-\alpha)}. \qquad (2.17)$$

At the same time, in the short-wave part of the radio band due to the large noise of receivers, masers considerably increase the sensitivity of the device as a whole despite the possible increase in atmospheric noise. In fact, for large values of F formula (2.16) reduces to

$$q = \frac{1-\alpha}{1-\alpha_0} G. \qquad (2.18)$$

Hence the use of masers as input stages of receivers intended for the reception of regular signals considerably improves the sensitivity. In the case when using expression (2.17) this improvement is due to the elimination of the effect of the inherent noise of the amplifier so completely that the threshold of sensitivity is determined by the level of the external noise in relation to the amplifier noise. In the case when formula (2.18) is used the improvement in sensitivity is due to the effective reduction in the receiver noise by a factor G.

Note that an improvement in sensitivity by a factor q leads to an increase in the transmission range of a communication link by a factor of $q^{1/2}$, and of radar stations by a factor of $q^{1/4}$.

Analysis of expression (2.16) for q enables us to formulate the requirements as regards the antenna-feeder channel of the receiving system which must be satisfied in order to attain a high gain in sensitivity. The chief of these is the need for a sharp reduction in the loss of microwave energy. The invention of masers necessitates a review of the level of this requirement, which was established before masers appeared. At the present time antenna-feeder systems are not regarded as good enough if the loss in energy in them exceeds 1-2dB. An important characteristic of an antenna system from the point of view of obtaining high gain is also the absence of side and backward lobes of the directional pattern, which can receive thermal radiation from the earth's surface.

We will now consider the increase in capacity of the communication channel when using masers as the input stages of the receivers. As is well known, the capacity of a communication channel (in binary units) is

$$C = \Delta\nu \log_2(1 + E_{sig}/kT_{in}\Delta\nu), \qquad (2.19)$$

where E_{sig} is the intensity of the signal which carries the information. If the passband of the communication channel is not changed when the maser is connected in the communication link, the capacity of the channel increases in accordance with formula (2.19) in proportion to the reduction of T_{in} when the maser is connected. In the case of small signals the quantity C increases by a factor q. However, the problem of the information capacity of communication channels when using masers in the receiver systems of the communication lines requires special consideration. This capacity increases not only when T_{in} is reduced, but also when the reception bandwidth is increased. However, the use of masers while simultaneously reducing the effective temperature of the input noise may lead to a reduction in $\Delta\nu$, since T_{in} is reduced when the gain G is increased, and for masers the attainable gain is always rigidly connected with the passband which can be realized in this case. Thus, for a single-cavity maser the product $G^{1/2}\Delta\nu$ is constant, for a two-cavity maser the product $G^{1/4}\Delta\nu$ is constant, and for a traveling-wave maser the product $\Delta\nu\sqrt{\ln G}$ is constant. Consequently, for each type of amplifier there is a gain providing the largest channel capacity. This optimum gain is realized if the width of the spectrum of the signal which carries the information E_{sig} is chosen equal to the passband of the maser $\Delta\nu$. However, in practice it is technically more convenient to use masers in communication links only to ensure the required small level of input noise in the given passband than to also try to optimize the communication system as a whole.

Devices which receive noisy signals, for example, radio astronomy receivers, require similar treatment. It follows from formula (1.9) that the least distinguishable change in temperature of the input noise δT for the radio band equals

$$\delta T = T_{in}\sqrt{1/\Delta\nu\tau}. \tag{2.20}$$

It is well known that the sensitivity of such receivers is higher the wider the reception band. Hence, when considering the problem of the sensitivity of radiometers with masers we must also bear in mind the relations which connect the gain and the bandwidth. For amplitude radiometers which receive signals with a continuous spectrum such a consideration is of particular interest in designing the optimal constructions. For frequency radiometers, intended for the reception of radiation in the spectral line of cosmic radio radiation, the gain in sensitivity is q times the gain in sensitivity when receiving regular signals.

It is quite easy to make a simple estimate of the value of the optimal amplification and the gain in sensitivity corresponding to it. For a single-cavity maser for which the quantity

$$\Delta\nu G^{1/2} = A \tag{2.21}$$

is constant, expression (2.20) can be written in the form

$$\delta T = G^{1/4}\left(T_m + \frac{F-1}{G}T_s\right)\sqrt{1/\tau A}, \tag{2.22}$$

where $T_m = T_b + \frac{\alpha}{1-\alpha}T_t + \frac{|T_{eff}|}{1-\alpha}$ [see formula (2.13)]. It is easy to see that when

$$G_{opt} = 3(F-1)T_s/T_m \tag{2.23}$$

the best possible sensitivity is obtained, viz.,

$$\delta T_m \approx (F-1)^{1/4}T_s^{1/4}T_m^{3/4}\sqrt{1/\tau A}. \tag{2.24}$$

In this case the gain in sensitivity with respect to a radiometer without a maser is

$$q_{opt} \approx (F-1)^{3/4}\left(\frac{T_s}{T_m}\right)^{3/4}\sqrt{A/\Delta\nu_0}, \tag{2.25}$$

where $\Delta\nu_0$ is the bandwidth of the linear path of the receiver of a radiometer without a maser. When operating with a low-temperature source and for good channels this reduces to

$$q_{\text{opt}} \approx \left(\frac{F-1}{\alpha}\right)^{3/4} \sqrt{A/\Delta v_0}. \tag{2.25a}$$

Similarly, for a two-resonator maser

$$G_{\text{opt}} \approx 7(F-1)\frac{T_s}{T_m}. \tag{2.26}$$

Consequently,

$$q_{\text{opt}} \approx (F-1)^{7/8}\left(\frac{T_s}{T_m}\right)^{7/8}\sqrt{A/\Delta v_0} \tag{2.27}$$

and for the particular case already considered

$$q_{\text{opt}} \approx \left(\frac{F-1}{\alpha}\right)^{7/8}\sqrt{A/\Delta v_0} \tag{2.27a}$$

a more thorough analysis, carried out by us in [15], taking into account the shape of the frequency characteristics of the single-cavity amplifier and the intermediate-frequency amplifier of the radiometer, leads to approximately the same results. In this case it turns out that there exists a weakly expressed optimum as regards the passband of the intermediate-frequency amplifier and values of the optimal amplification are obtained which differ somewhat from expression (2.23). Thus, for a rectangular shape of the passband of the intermediate-frequency amplifier we obtain $G_{\text{opt}} = 4.5(F-1)T_s/T_m$. The expression for the gain (2.25) in this case agrees with those given in [15].

The optimal values of the gain (2.25) and (2.27) can only be realized in the decimeter and centimeter waveband. In the very short wavelength part of the radio band where F is large the optimal amplification cannot be attained since cavity masers with amplifications considerably exceeding 20 dB do not operate sufficiently stably. When the optimum cannot be attained due to very large F, the gain in sensitivity of amplitude radiometers in the case when single and two-cavity masers are used is

$$q = G^{3/4}\sqrt{A/\Delta v_0}, \tag{2.28}$$
$$q = G^{7/8}\sqrt{A/\Delta v_0} \tag{2.29}$$

respectively. From these relations we see the advantage of two-cavity masers.

Therefore, in radio astronomy receivers the use of cavity masers can lead to a considerable gain in sensitivity, if the constant A (see Chapter IV), which characterizes the effectiveness of the active material of the maser is not less than the bandwidth of the receiver used in the radiometer without masers. In the decimeter and centimeter bands for losses in the channel of 0.5 dB ($\alpha = 0.1$) and $F-1 \approx 10$ for $A = \Delta v_0$ the maximum gain for a single-cavity maser is 30 and for a two-cavity maser 55.

Multicavity masers and traveling-wave masers must be considered separately, as by using the method developed in [15]. If they are fairly broadband, formulas (2.15)-(2.18) apply for estimating the gain in sensitivity. It should be borne in mind that in traveling-wave masers the nonreciprocal element is cooled and we can therefore assume that in practice $\alpha = \alpha_0$.

§ 3. Sensitivity of Radio Receivers with a Noncoherent Intensity Detector (Square-Law Video Reception)

The development of quantum electronics, the construction of generators of monochromatic oscillations, and the spread of radiophysical methods of investigation over the whole of the very-high-frequency band (the submillimeter and optical bands) — all this underlines the fundamental significance and practical importance of the use of masers. The use of linear amplifiers enables us, by reducing the passband of the linear part of the receiver, to improve its sensitivity considerably when receiving monochromatic signals, and also to employ phase methods which have been well developed in radio engineering.

However, in certain cases the noncoherent recording of the intensity of the received signal may be quite sufficient. This includes many problems in radio astronomy, passive radar, plasma physics, low-resolution spectroscopy for thermal sources of radiation in the submillimeter band, etc. At the same time, as was shown in Chapter I, the limiting sensitivity of noncoherent detectors of intensity is equivalent to the recording of a single quantum of radiation in the time of observation. Unfortunately, in the radio and submillimeter band at the present time there are no detectors whose sensitivity is near to this limit. The highest sensitivity is now obtained using Putley receiving elements [19], which are crystals of n-type InSb cooled to liquid-helium temperatures. With the proper choice of the density and mobility of the impurities these crystals are broadband photoresistors in the microwave band.

We will consider the sensitivity of radio receivers with noncoherent detectors of intensity under the conditions when the spectral density of the noise at the detector input is determined by a relation similar to formula (2.11). We will consider square-law video detection, since when measuring the intensity of signals with a continuous spectrum the operation of squaring the fluctuation voltages (the voltage of the fluctuation field) minimizes the spread of the output readings irrespective of the dependence on the shape of the input spectrum and the averaging law at the output [20, 21]. In addition, it is necessary to bear in mind that all thermal devices — bolometers, photoresistors, etc. — are square-law.

Thus, we will consider a square-law detector. At its input within the passband $\Delta \nu$ for N types of oscillations is fed a noise spectral density

$$P_\nu = ((1-\alpha)\rho_b + \alpha \rho_t) N, \qquad (2.30)$$

where we have introduced the notation

$$\rho_b = \frac{h\nu}{e^{h\nu/kT_b} - 1} \quad \text{and} \quad \rho_t = \frac{h\nu}{e^{h\nu/kT_t} - 1}.$$

Formula (2.30) differs from formula (2.11) in not containing the spectral density of the thermal noise of the detector at high frequency, since the detector does not rectify the inherent thermal noise (see, for example, [22]). We will assume that the spectral densities ρ_b and ρ_t exceed the energy of a quantum $h\nu$. Then for video frequencies $\nu_b < \Delta \nu$ the spectral density of the intensity of the fluctuations of the output voltage, due to the noise P_ν, is uniform and equals

$$S_1 = K^2((1-\alpha)\rho_b + \alpha \rho_t)^2 N \Delta \nu, \qquad (2.31)$$

where K is the slope of the curve of the power converted at the input of the detector to the voltage at the output (at the video frequency). To this we must add the spectral densities of the intensity of the voltage fluctuations at the video frequency, due to the low-frequency noise of the detector and the noise of the measuring video amplifier, at the input of which the detector is connected. If the output resistance of the detector (the resistance at the video frequency) equals R_d, the input resistance of the amplifier equals R_{in}, and its equivalent noise resistance equals R_n, then when $R_{in} \gg R_d$ the spectral density of the noise at the input of the amplifier equals

$$S = K^2[(1-\alpha)\rho_b + \alpha \rho_t]^2 N \Delta \nu + 4R_d k T_d + 4R_n k T_n, \qquad (2.32)$$

where T_d is the temperature of the detector and T_n is the temperature connected with the equivalent noise resistance R_n. Usually $T_n = 300°K$. We will assume that there is no excess noise in the detector (shot noise or current noise).

With the help of formula (2.32) it is easy to obtain an expression for the sensitivity in relation to a monochromatic signal

$$E = \sqrt{\frac{\left(\rho_b + \frac{\alpha}{1-\alpha}\rho_t\right)^2 N \Delta \nu}{\tau} + \frac{4R_d k T_d + 4R_n k T_n}{K^2(1-\alpha)^2 \tau}}, \qquad (2.33)$$

where, as usual, the time constant of the output device $\tau > 1/\Delta\nu$. Similarly, the minimum detectable change in the energy of a single oscillator of the field of the background thermal radiation $\Delta\rho_b$ equals

$$\Delta\rho_b = \sqrt{\frac{\left(\rho_b + \frac{\alpha}{1-\alpha}\rho_t\right)^2}{N\Delta\nu\tau} + \frac{4R_d k T_d + 4R_n k T_n}{K^2(1-\alpha)^2 N^2 \Delta\nu^2 \tau}}. \qquad (2.34)$$

The relations obtained differ from formulas (1.11) and (1.12), which describe the limiting sensitivity of noncoherent receivers of radiation, first of all by the presence of a second term, which vanishes as the slope of the detector conversion curve K increases. Under practical conditions, however, K is small and formulas (2.33) and (2.34) must be written in the form

$$E = \frac{1}{K(1-\alpha)}\sqrt{\frac{4R_d k T_d + 4R_n k T_n}{\tau}}, \qquad (2.35)$$

$$\Delta\rho_b = \frac{1}{K(1-\alpha)N\Delta\nu}\sqrt{\frac{4R_d k T_d + 4R_n k T_n}{\tau}}. \qquad (2.36)$$

It is obvious that the construction of a video amplifier must ensure that the condition $R_n k T_n \ll R_d k T_d$ is satisfied. If the detector used in reception has a small slope the level of the output fluctuations is determined not by the detected noise but by the noise at the video frequency. This, generally speaking, trivial fact leads to an essentially different dependence of the sensitivity of receivers on the bandwidth of the received frequencies than in the case when using linear amplifiers. When receiving noisy signals it can happen that the video receiver is more sensitive than a receiver with a maser, since the reception bandwidth for the detector is considerably larger than for the amplifier. However, it is not easy to obtain this advantage at the present time. Thus, for $K = 10^3$ V/W, $R_d = 100\ \Omega$, and $T_d = 4.3°K$ at a wavelength of the order of 1 cm over a 10% bandwidth $\Delta\nu = 3\cdot 10^9$ Hz for losses in the channel $\alpha = 0.5$, the temperature sensitivity of the detector for a time constant of 1 second is 8°K. At the same time a maser with a bandwidth of $3\cdot 10^7$ Hz for $\alpha = 0.5$, a channel temperature $T_t = 300°K$ and a background temperature $T_b = 150°K$, has a sensitivity of 0.1°K for a time constant of 1 second.

At the present time, obviously, only in those receivers in which linear amplifiers are used is the sensitivity determined by the noise at the high frequency. In the case when masers are used this noise is the input noise, which is external in relation to the amplifier. Nevertheless, video-detector receivers are of great interest particularly in those bands for which there are no masers.

§4. The Technical Threshold of Sensitivity and the System Requirements when Using Masers

The relations obtained in the previous section show that the large increase in sensitivity which is possible when using masers or high-sensitivity detector elements in radio receivers is obtained when the antenna-feeder channel of the device has small losses, and the directional pattern of the antenna does not have any back or side lobes. This dictates obvious requirements as regards the construction of the receiver systems. The relations derived above for the sensitivity of different types of receivers enable us to choose under different concrete conditions the optimum solutions for which the maximum possible gain in sensitivity is obtained. In particularly important cases to obtain a high gain in sensitivity it is necessary to cool the dissipative elements of the input channel, such as the decoupling isolators, the modulators, the switches, etc.

In addition, it should be borne in mind that the formulas derived earlier relate to the natural threshold of sensitivity of radio receivers. Under practical conditions the technical threshold of sensitivity has a large value, due to the nonideal nature of the apparatus and the meagerness of our practical skill in stabilizing its parameters or eliminating the effect of variations in them, whereas the natural threshold depends on fundamental physical effects such as thermal noise, spontaneous noise, the shot effect, etc.

The use of masers, which considerably lower the natural threshold of sensitivity, increases the requirements as regards the stability of the receiving apparatus, since to fully realize the full possibilities of ideal receiver elements requires adequate improvement of the whole receiver channel. In this case, whereas a reduction in the losses in the antenna-feeder channel improves the natural threshold of sensitivity, a reduction in the matching of the channels and the stabilization of their electrical lengths raises the technical threshold of sensitivity. Moreover, the technical threshold of sensitivity is also determined by the quality of the electronic units of the receiver (the amplifier stability, the heterodyne frequency, etc.)

In describing the technical threshold of sensitivity we will assume that the parameters of the receiving apparatus undergo both stationary and relatively fast fluctuations, and also slow and nonstationary changes. Having in mind the apparatus used in radio astronomy, we will assume that the correlation time τ_f of the stationary fluctuations in the parameters $f(t)$ is less than the time constant of the output device τ. Therefore, the contribution of these fluctuations to the general level of the output noise does not differ in structure from the fluctuations due to internal noise. At the same time the slow drifts, varying from day to day and being by nature irregular, come out during concrete measurements as almost regular and ideally unaveraged due to their slowness. The effect of such nonstationary drifts is stronger the longer the time of observation. The time of observation for the most part depends on the nature of the processes being investigated. In radio astronomy it is the time taken by a source to pass through the directional pattern of the radio telescope or the time of obscuration of the source by some celestial body, etc. It is obvious that the observation time θ always exceeds the time constant of the output device τ.

The presence of zero drifts which are nonstationary during observation can be ascribed to the effect of extremely slow stationary fluctuations $S(t)$, the correlation time of which τ_S considerably exceeds the observation time θ.

Therefore, the times τ_f, τ, τ_S and θ, which are characteristic of the technical threshold of sensitivity, are connected by the inequalities

$$\tau_f \ll \tau \ll \theta \ll \tau_S. \tag{2.37}$$

It is obvious that the fluctuations of $f(t)$ and $S(t)$ are quasistationary, similar to slow modulation, and pass through the whole channel of the receiver which precedes the output device.

When using practical receivers the problem of eliminating fluctuations of the parameters cannot be solved merely by means of high stabilization. It is necessary therefore to use methods of operation for which the presence of instabilities has the least effect. All kinds of modulation methods have become widely used. For a modulation frequency which exceeds the width of the spectrum of the parameter fluctuations in fact, as is well known, it is possible to filter out a considerable part of the excess noise, the spectral density of which approaches zero. We will consider the residual effect of parameter fluctuations for the modulation method of reception [23]. With the correct choice of the modulation frequency $\Omega \gg 1/\tau_f$ the effect of the fluctuations of $f(t)$ and $S(t)$ appears only when there is a modulated signal present. The natural threshold of sensitivity in this case is hardly changed. Fluctuations due to random modulation of the signal of frequency Ω by the processes $f(t) + S(t)$ are added to the spectral density of the output noise. These additional fluctuations are determined both by the intensity of the processes $f(t)$ and $S(t)$, and by the value of the signal of frequency Ω, which is formed when detecting the high-frequency processes modulated at a frequency Ω.

If the modulation is subject to a power $\Delta P_\nu \Delta \nu$, then at the input of the output device at the modulation frequency there is fed a voltage $G \Delta P_\nu \Delta \nu$, where G has the meaning of the over-all gain of the complete receiver−amplifier channel as a whole. We introduce the parameter fluctuations by representing G in the form $G = G_0[1 + f(t) + S(t)]$. Then the relative intensity of the output fluctuations which correspond to the rapid fluctuations of $f(t)$ is $\overline{f^2}(\Delta P_\nu \Delta \nu)^2 \frac{\tau_f}{\tau}$, which gives the value of the stationary part of the technical threshold.

To estimate the nonstationary part we can assume that the root-mean-square drift in a time θ equals the product of a certain effective drift speed $\frac{1}{\tau_S}\sqrt{\overline{S^2}} = \sqrt{(\dot{S})^2}$ by the time θ [24]. We then obtain for the nonstationary part of the threshold the quantity $(\Delta P \Delta \nu)^2 \overline{(\dot{S})^2} \theta^2$. Taking all these estimates into account formula (2.20) can be written in the form

$$\delta T = \left[T_{in}^2 \frac{1}{\Delta \nu \tau} + \Delta T^2 \left(\overline{f^2} \frac{\tau_f}{\tau} + (\overline{\dot{S}})^2 \theta^2 \right) \right]^{1/2}. \qquad (2.38)$$

We see from the expression obtained how the presence of a technical threshold of sensitivity sets a limit to the improvement in sensitivity which is possible by reducing the level of the input noise of the receiver and broadening the reception passband if one does not ensure a corresponding increase in stability.

We note here that in those cases when T_{in} is determined by the presence of an external medium (the atmosphere), the problem of the stability of this radiation becomes important. The effect of such instabilities can be reduced by using different correlation methods, by considerably reducing the observation time, etc.

The presence of a nonzero modulation signal ΔT may be due to a number of causes. In radio astronomy observations it is sometimes difficult to obtain the same background radiation in two positions of the input modulator (switch).

Hence, the presence of fluctuations in amplification can lead to a considerable increase in the technical threshold of sensitivity of high-sensitivity radio receivers. Modulation methods of reception provide a considerable but not complete suppression of the effects of these fluctuations. A further reduction of this effect can be obtained in radio astronomy by the use of automatic null radiometers [25], a theoretical [26] and experimental [27] investigation of which indicates the possibility of obtaining the natural threshold of sensitivity. Radiometers of this type are tracking systems. The voltage from the output of the receiver, the input of which is periodically switched from the investigated source to a noisy equivalent, is used to control the effective temperature of the noise of the equivalent source, so that the output signal is always close to zero. If T_{in} is reduced, the requirement as regards the accuracy of operation of this tracking system increases. In certain cases for one technical reason or another it is not possible to construct an automatic null radiator.

The use of automatic gain control (AGC) by the level of inherent noise, in order to reduce the technical threshold of sensitivity of radiometers by reducing the effect of gain fluctuations, can lead to an increase in the effect of noise factor fluctuations [28], which can be suppressed in modulation radiometers by a correct choice of the modulation frequency. Experiments on masers in radiometers at wavelengths of 21 cm [29], have verified this theoretical prediction.

We point out once again the simple conclusion that a reduction in the natural threshold of sensitivity of receivers requires an adequate improvement in the stability of operation of all the sections of the receiver-amplifier channels. The problem of the stability of the amplitude and phase characteristics of the masers proper will be considered in the following sections, devoted directly to masers.

In connection with the discussion concerning the question of the suppression of the parasitic modulation signal ΔT we note that the use of nonreciprocal elements in the input channels of the radio telescope practically completely removes interference by noise due to the inherent thermal radiation of receivers in the case of sufficiently well matched antenna-feeder channels of finite lengths [30, 31]. However, even if the channel contains an isolator with ideal decoupling, the parasitic modulation signal is not completely suppressed [31]. A unidirectional isolator of any form necessarily contains a completely absorbing attenuator. Its thermal radiation propagates into the antenna channels, is partially reflected, and in view of the difference in reflection coefficients, in this case is modulated. After reflection this radiation propagates as a signal from the antenna to the receiver and therefore passes through the unidirectional isolator. Only strong cooling of the absorbing element of the isolator can completely eliminate this parasitic modulation. We recall that cooling of the dissipative elements of the input channel is advisable in order to obtain a large gain in sensitivity.

The problem of the technical threshold of sensitivity also includes the noise immunity of the receiver. It is clear that a reduction in the level of the input noise increases the requirement as regards the noise protection of the receiver. In this case, for receivers with a very broad band of received frequencies, the danger of receiving interference signals is very large. For video receivers, particularly those whose sensitivity is described by formulas (2.35) and (2.36), not only the high-frequency noise, but also the noise at the frequencies of the video band are important. However, in this case reliable screening of the receiver is possible. In addition, if a modulation method is used the criterion of the correctness of the choice of modulation frequency is the tuning out not only of the excess noise but also of the noise at the video frequency, such as network harmonics, etc. To achieve this, the receiving element of the video receiver should have very little inertia.

CHAPTER III

The Q-Factor of the Active Material

To determine the parameters of masers such as the gain, the bandwidth, the stability, and the gain stability, etc., in each concrete case we can solve the straightforward problem of the interaction of the electromagnetic field in some electrodynamic system or other (waveguide, resonator, etc.) with the active material. However, this would lead to very formidable and often quite impracticable calculations. Nevertheless, in developing the theory of masers it is possible to divide the consideration of its properties into two parts. In the first stage we determine and calculate the Q-factor of the active material which characterizes its amplifying properties. In the second stage this Q-factor is used in the usual relations of circuit theory and enables us to obtain the parameters of the different types of masers.

In this chapter we use the density matrix method to calculate the Q-factor of the active material of paramagnetic masers [32] with three energy levels and two relaxation times. By taking into account the detuning of the pumping radiation we have been able to determine the phase stability of the Q-factor of the active material.

§1. Q-Factor and Nondiagonal Elements of the Density Matrix

The negative nature of the real part of the complex Q-factor of the active material corresponds to the presence of negative losses of energy when there is population inversion. We will consider for definiteness the Q-factor of the active material placed in a cavity resonator. By definition, the Q-factor of the material is

$$Q_\text{в} \equiv \frac{2\pi \nu E_\text{c}}{P_\text{c}}, \qquad (3.1)$$

where ν is the center frequency, E_c is the energy stored in the resonator, and P_c is the power dissipated in the material. When there is population inversion the power dissipated in the material is negative since the material radiates and does not absorb energy. The radiated power equals the product of the energy of a quantum times the difference in the numbers of radiated and absorbed particles and also times the probability of induced radiation in unit time

$$P_\text{rad} = -P_\text{c} = I \Delta n_0 h\nu W_\text{c}, \qquad (3.2)$$

where W_c is the probability of transitions induced by the signal, Δn_0 is the equilibrium difference in populations of the signal transition, and I is the inversion coefficient which denotes the ratio of the excess number of particles in the upper level when there is inversion to the excess number of particles in the lower level when there is thermal equilibrium. In the radio band and in the case when the system has Z levels

$$P_\text{rad} = \frac{NI}{ZkT}(h\nu)^2 W_\text{c}, \qquad (3.3)$$

where N is the total number of particles in all the Z levels.

The value of the radiated power characterizes the amplifying properties of the active material. This quantity changes when the frequency of the signal changes in accordance with the frequency dependence given by the form factor of the line $g(\nu)$, which is proportional to the probability W_c. Formulas (3.1)-(3.3) enable us to estimate the Q-factor of the active material and to obtain its dependence on the signal frequency. However, in such an estimate we cannot determine the dispersion properties of the Q-factor of the material and we cannot take into account the effect of the pumping frequency. The Q-factor of the active material can be found more accurately by calculating its complex susceptibility. We will determine first of all the connection between the Q-factor and the susceptibility of the material (see, for example, [33]).

$$1/Q_\text{в} = j\eta\chi, \qquad (3.4)$$

where χ is the susceptibility, and the filling factor η determines the efficiency with which the high-frequency field is used and equals the ratio of the energy stored in the volume of the active material to the energy stored in the whole resonator. If the filling factor is known the problem of determining the Q-factor of the active material of paramagnetic masers reduces to the problem of determining the complex magnetic susceptibility at the signal frequency, which is usually represented in the form $\chi = \chi' - j\chi''$. The real part of the susceptibility specifies the reactive effect of the material, and its imaginary part determines the gain. When $\chi'' < 0$ the real part of the Q-factor is negative and an amplification effect is obtained. The susceptibility can be obtained knowing the polarization of the material. In turn, the polarization is determined by the matrix elements of the operator of the dipole moment μ_{ik} and by the elements of the density matrix ρ_{ik}. In the case of N particles in unit volume the polarization of the material M is given by

$$\mathbf{M} = N \sum_{ik} \rho_{ki} \boldsymbol{\mu}_{ik}. \qquad (3.5)$$

For three-level masers with the signal transition 3 → 2 the susceptibility at the signal frequency can be obtained if the polarization components at that frequency are known.

$$M_{32} = N(\rho_{23}\mu_{32} + \rho_{32}\mu_{23}) = 2N \operatorname{Re}(\rho_{23}\mu_{32}). \tag{3.6}$$

Therefore, the problem of determining the complex Q-factor of the active material of a three-level maser reduces to calculating the nondiagonal element of the density matrix $\rho_{23} = \rho_{32}^*$, which corresponds to the signal transition.

The density matrix was used by Kontorovich and Prokhorov [34] and by Anderson [35] to describe a three-level system situated in a strong pumping field and in a weak signal field. In [34], by considering a gas with one relaxation time, the authors arrived at a conclusion concerning the splitting of the generation lines for a very strong pumping field and concerning the dependence of the generation frequency on the frequency and amplitude of this field. In [35], also assuming the existence of only one relaxation time, the conditions for the excitation of oscillations were analyzed for the rather special case of a small change of the equilibrium populations of the levels. This paper is interesting, however, in the fact that it introduces a density matrix element ρ_{12}, which corresponds to a free transition, and takes into account the reaction of the field excited by the polarization component M_{12}. Butcher [36] calculated the nondiagonal element ρ_{23}, neglecting the presence of ρ_{12}. In the literature on masers a formula for the Q-factor of the active material is widely used based on the results given in [36] (see, for example, [33]).

In accordance with the results given in [33, 36] the phase and amplitude of the Q-factor of the material do not depend on the intensity and frequency of the pumping radiation when the pumping transition is saturated. This stability of masers is very important and hence a more rigorous consideration which enables us to estimate more carefully the effect of the pumping radiation is of considerable interest.

The equations of motion of the density matrix as they apply to the three-level paramagnetic maser have been solved by Clogston [37], taking into account the nondiagonal element ρ_{21} and the reverse reaction of the field on the frequency of the free transition ν_{21}. However, Clogston did not take into account the possibility of the detuning of the pumping radiation and essentially did not obtain explicit expressions for the Q-factor of the active material under conditions of practical interest. By taking into account the possibility of detuning we are able to obtain an expression for the Q-factor of the active material and to determine the effect of changes in frequency and intensity of the pumping on the real and imaginary part of the Q-factor Q_B, which governs the phase and amplitude stability of the masers. We will assume in what follows that the line is uniformly broad.

§2. Calculation of the Susceptibility

We will introduce the following notation. Let $(E_n - E_m)/\hbar$ be the frequency of the transition between the energy levels E_n and E_m; Ω_{nm} is the resonance frequency of the magnetic field with amplitude $H_{mn} = H_{nm}$ and phase $\varphi_{mn} = -\varphi_{nm}$; $\tau_{nm} = \tau_{mn}$ is the relaxation time of the nondiagonal element of the density matrix (the spin-spin relaxation time) and $w_{kn} \neq w_{nk}$ is the rate of relaxation of the diagonal elements (the probability of spin-lattice relaxation). Then the equations of motions of the density matrix can be written in the form

$$\begin{aligned}\dot{\rho}_{nn} &= \sum_k (\omega_{kn}\rho_{kk} - w_{nk}\rho_{nn}) + \frac{1}{j\hbar}\sum_k(V_{nk}\rho_{kn} - \rho_{nk}V_{kn}), \\ \dot{\rho}_{nm} &= \frac{1}{\tau_{nm}}\rho_{nm} + \frac{1}{j}\omega_{nm}\rho_{nm} + \frac{1}{j\hbar}\sum_k(V_{nk}\rho_{km} - \rho_{nk}V_{km}),\end{aligned} \tag{3.7}$$

where the matrix element of the excitation operator is

$$V_{nm} = -\mu_{nm} H_{nm} \cos(\Omega_{nm} t - \varphi_{nm}). \tag{3.8}$$

The substitution

$$\rho_{nm} = \sigma_{nm} e^{-j\omega_{nm} t} \tag{3.9}$$

leads to the concept of interaction, and the substitution

$$A_{nk} = \frac{1}{2\hbar} H_{nk} \mu_{nk} e^{j\varphi_{nk}}, \tag{3.10}$$

$$\sigma_{nk} = \lambda_{nk} e^{-j(\Omega_{nk} - \omega_{nk})t}$$

shortens the notation and enables us to introduce the detuning between the frequencies of the fields and of the transition

$$\delta_{nk} = \Omega_{nk} - \omega_{nk}. \tag{3.11}$$

As a result, for stationary conditions when $\dot{\lambda}_{nm} = 0$ we obtain the equations

$$\sum_k (w_{kn}\lambda_{kk} - w_{nk}\lambda_{nn}) = -j\sum_k (A_{nk}\lambda_{kn} - \lambda_{nk}A_{kn}), \tag{3.12}$$

$$(-j\delta_{nm} + 1/\tau_{nm})\lambda_{nn} = j\sum_k (A_{nk}\lambda_{km} - \lambda_{nk}A_{km}).$$

The equality $\dot{\lambda}_{nm} = 0$ indicates that under stationary conditions the amplitude of the oscillations of each nondiagonal element of the density matrix is constant, and the frequency is the same as the frequency of the corresponding stimulating field Ω_{nm}. When solving equations (3.12) it must be borne in mind that for a three-level system $\omega_{31} = \omega_{32} + \omega_{21}$ and $\Omega_{31} = \Omega_{32} + \Omega_{21}$. In addition, $\rho_{kk} = \lambda_{kk}$ mean the relative populations and are subject to the normalization condition

$$\rho_{11} + \rho_{22} + \rho_{33} = 1. \tag{3.13}$$

It is difficult to solve equations (3.12) in general form. Therefore, in what follows it will be assumed that the polarization of the material at the frequency of the pumping radiation, i.e., the nondiagonal element ρ_{31}, and the distribution of the populations between the levels, i.e., the diagonal elements ρ_{11}, ρ_{22}, and ρ_{33}, are determined only by the speeds of the corresponding relaxation processes and by the strong field of the pumping radiation A_{13}. In this case at the next stage of the consideration for determining the nondiagonal elements ρ_{32} and ρ_{21}, the accurate equations are used in which the mutual effect of the components of the polarization at the frequencies Ω_{32} and Ω_{21} is taken into account.

In accordance with what has been stated above, from equations (3.13) we separate the equations

$$\left(j\delta_{31} + \frac{1}{\tau_{31}}\right)\lambda_{13} = jA_{13}(\rho_{33} - \rho_{11}) \tag{3.14}$$

and

$$-(\omega_{12} + w_{13})\rho_{11} + w_{21}\rho_{22} + w_{31}\rho_{33} = -j(A_{13}\lambda_{31} - \lambda_{13}A_{31}),$$

$$w_{12}\rho_{11} - (w_{21} + w_{23})\rho_{22} + w_{32}\rho_{33} = 0. \tag{3.15}$$

Equations (3.15) taking into account the expressions (3.14) and (3.13) are the usual rate equations for a three-level maser and are easily solved. For relative population differences $\Delta_{31} = \rho_{33} - \rho_{11}$ and $\Delta_{32} = \rho_{33} - \rho_{22}$ we obtain

$$\Delta_{31} = -\frac{\Delta_{13}^0}{1 + \dfrac{T_1 \tau_{31}}{1 + \delta_{31}^2 \tau_{31}^2} |A_{13}|^2} \qquad (3.16)$$

and

$$\Delta_{32} = \frac{-\Delta_{23}^0 + \dfrac{T_1' \tau_{31}}{1 + \delta_{31}^2 \tau_{31}^2} |A_{13}|^2}{1 + \dfrac{T_1 \tau_{31}}{1 + \delta_{31}^2 \tau_{31}^2} |A_{13}|^2}, \qquad (3.17)$$

where Δ_{13}^0 and Δ_{23}^0 are the equilibrium population differences of the levels 1-3 and 2-3 respectively, and the effective time T_1 is determined by the probabilities of spin-lattice relaxation processes

$$T_1 = (2w_{21} + w_{12} + w_{32} + 2w_{23})[(2w_{21} + w_{12} + w_{32} + 2w_{23})(w_{21} + 2w_{12} + w_{31} + 2w_{13}) +$$
$$+ (2w_{21} + w_{12} + w_{31} + w_{13})(w_{21} - 2w_{12} + w_{32} + w_{23})]^{-1}. \qquad (3.18)$$

In this case the ratio

$$\frac{T_1'}{T_1} = \frac{w_{21} - w_{12} - w_{32} + w_{23}}{2w_{21} + w_{12} + w_{32} + 2w_{23}} \qquad (3.19)$$

gives the value of the difference in the relative populations of the signal transition, ultimately reached in a system of three energy levels when the pumping level is saturated. The condition of saturation is the satisfaction of the inequality

$$\frac{|A_{13}|^2 T_1 \tau_{31}}{1 + \delta_{31}^2 \tau_{31}^2} \gg 1. \qquad (3.20)$$

In what follows we will assume that inversion is attained for which Δ_{32} exceeds by a factor I the equilibrium population difference Δ_{23}^0,

$$\Delta_{32} = I \Delta_{23}^0, \qquad (3.21)$$

where I is the positive inversion coefficient. It is obvious that $\Delta_{21} = \Delta_{31} - \Delta_{32}$.

The inversion coefficient is a very important quantity which characterizes the efficiency of the system of levels used in the maser. As can be seen from formula (3.19), the value of the inversion coefficient is essentially determined by the relation between the probabilities of the relaxation transition and the frequencies. Without carrying out an analysis of the well-known properties of the inversion coefficient we will simply note that the quantity I can be related to the effective (negative) temperature T_{eff} introduced in Chapter I. In the presence of inversion I formula (1.23) for T_{eff} can be written in the form

$$T_{\text{eff}} = -\frac{h\nu_{ij}}{k} \frac{1}{\ln\left(1 + \dfrac{I_{ij}}{\rho_{ji}}\right)}, \qquad (3.22)$$

where j is the number of the lower level.

In a three-level system when $h\nu_{ij} \ll kT$, where T is the temperature of the thermostat, formula (3.22) reduces to the simple relation

$$T_{\text{eff}} = -T/I, \qquad (3.23)$$

which gives a clearer meaning to the negative effective temperature.

A knowledge of the differences in the population Δ_{ik} and of the nondiagonal element λ_{13} gives the right side of the equations for the nondiagonal elements λ_{23} and λ_{21}. Introducing for brevity the notation $\gamma_{ik} = j\delta_{ik} + 1/\tau_{ik}$, we obtain from (3.12) the equations

$$\gamma_{21}\lambda_{21} + jA_{31}\lambda_{23} = jA_{21}\Delta_{12} + jA_{23}\lambda_{31},$$
$$\gamma_{32}\lambda_{23} + jA_{13}\lambda_{21} = jA_{23}\Delta_{32} + jA_{21}\lambda_{13}. \tag{3.24}$$

We see from these equations how the pumping radiation (A_{31} and A_{13}) accomplishes cross interaction between the polarizations for the signal and free transitions. The solution of (3.24) has the form

$$\lambda_{23} = \frac{j\gamma_{21}(A_{23}\Delta_{32} + j\frac{\Delta_{31}}{\gamma_{31}}A_{21}A_{13}) + A_{13}(A_{21}\Delta_{12} - j\frac{\Delta_{31}}{\gamma_{13}}A_{23}A_{31})}{\gamma_{21}\gamma_{32} + |A_{13}|^2},$$

$$\lambda_{21} = \frac{j\gamma_{32}(A_{21}\Delta_{12} - j\frac{\Delta_{31}}{\gamma_{13}}A_{23}A_{31}) + A_{31}(A_{23}\Delta_{32} + j\frac{\Delta_{31}}{\gamma_{31}}A_{21}A_{13})}{\gamma_{21}\gamma_{32} + |A_{13}|^2}, \tag{3.25}$$

where Δ_{ik} are given by the solutions of (3.15). For $A_{21} = 0$, λ_{21} does not revert to zero. Therefore the argument used in [33] on the possibility of ignoring *a priori* the nondiagonal matrix element which corresponds to the free transition when there is no resonator reaction on the radiation of components of polarization at the frequency of the free transition generally speaking is untenable.

Therefore the amplitudes of the nondiagonal elements λ_{23} and λ_{21} are obtained as a function of the fields A_{23} and A_{21} which are at the frequencies of the transitions $3 \to 2$ and $2 \to 1$. But the field A_{23} is the field of the external signal introduced by us. As a result of the simultaneous action on the three-level system of the fields A_{23} and A_{31} a dipole moment arises proportional to λ_{21} and which precesses at the difference (free) frequency Ω_{21}. This moment radiates a field of frequency Ω_{21}, which occurs in the equation for the density matrix and the solution (3.25) for λ_{23} and λ_{21}. The answer to the question of what field A_{21} is radiated by the moment λ_{21}, is determined by the value of the matrix element of the operator of the dipole moment μ_{21} and by the electrodynamic conditions of the radiation at the frequency Ω_{21}. It is obvious that there is a direct proportionality between the dipole moments and the fields. Introducing the coefficient of proportionality $\varkappa_{21}$ and $\varkappa_{23}$ by the relations

$$\lambda_{21} = \varkappa_{21}A_{21} \quad \text{and} \quad \lambda_{23} = \varkappa_{23}A_{23} \tag{3.26}$$

and substituting this expression in equation (3.25) we obtain uniform equations relative to A_{21} and A_{23}. The introduction of the coefficients $\varkappa_{21}$ and $\varkappa_{23}$ is equivalent to the statement of the proportionality of the field A_{21} and the field A_{23}. The condition of consistency of the equations for A_{21} and A_{23} gives the relation between the quantities $\varkappa_{23}$ and $\varkappa_{21}$:

$$\varkappa_{23} = \frac{|A_{13}|^2\left(\Delta_{32} + \frac{\Delta_{31}\gamma_{32}}{\gamma_{13}}\right)\left(\Delta_{12} - \frac{\Delta_{31}\gamma_{21}}{\gamma_{31}}\right)}{(\gamma_{21}\gamma_{32}+|A_{13}|^2)[\varkappa_{21}(\gamma_{21}\gamma_{32}+|A|^2)] - j\gamma_{32}\Delta_{12} - j\frac{\Delta_{31}}{\gamma_{31}}|A_{13}|^2} + \frac{j\gamma_{21}\Delta_{32} - j\frac{\Delta_{31}}{\gamma_{13}}|A_{13}|^2}{\gamma_{21}\gamma_{32} + |A_{13}|^2}, \tag{3.27}$$

where $\varkappa_{21}$ describes the radiational properties of the components of the polarization of the operating material at the free frequency.

Taking into account the definitions (3.26) and (3.10) and (3.9), we obtain

$$\rho_{23} = \frac{1}{2\hbar}\varkappa_{23}\mu_{23}H_{23}e^{-j(\Omega_{23}t - \varphi_{23})}. \tag{3.28}$$

Comparing expressions (3.28) and (3.6) with the definition of the susceptibility, we obtain

$$\chi_{23} = \frac{|\mu_{23}|}{\hbar}N\varkappa_{23}. \tag{3.29}$$

Formulas (3.29) and (3.27) determine the Q-factor of the active material in the most general case, although within the limits of the assumptions made for the separation of (3.12) in the rate equations (3.15) and the equations for the nondiagonal elements (3.14) and (3.24). In the case of accurate tuning of the frequency of the pumping radiation $\delta_{31} = 0$, and formula (3.27) reduces to the results obtained by Clogston [37].

§3. Q-Factor of the Active Material of a Paramagnetic Maser

The relations obtained in the previous section enable us to obtain the dependence of the real and imaginary parts of the Q-factor of the active material on the intensity and frequency of the pumping radiation in the general case. We see from equation (3.27) that for large intensities of the pumping radiation the susceptibility χ_{23} strongly depends on this intensity. In this case as $|A_{13}|^2$ increases the role of the radiation coefficient at the free frequency $\varkappa_{21}$ increases and the frequency dependence of the negative resistance at the signal frequency becomes double-humped.

We are interested in paramagnetic masers the operating material of which is characterized by two sharply different relaxation times T_1 and T_2. For usable materials $T_1 \approx 10^{-1} - 10^{-3}$ sec and $T_2 = 10^{-8}$ sec. The spin-lattice relaxation time T_1 is determined by the combination of partial probabilities of spin-lattice relaxation w_{ik} and in practice is the same as the effective relaxation time introduced in formula (3.18) to describe population inversion. The spin-spin relaxation time of any EPR lines T_2 is determined by the partial relaxation times of the nondiagonal elements of the density matrix τ_{ik} of the corresponding transitions. We will assume for simplicity that all the τ_{ik} are equal to one another and we will denote them by T_2. To obtain maximum inversion it is necessary to satisfy the condition (3.20) which implies that the probability of transitions induced by the pumping radiation field considerably exceeds the rate of spin-lattice relaxation. In view of the fact that the inequality $T_2 \ll T_1$ is satisfied with a large margin this probability, as a rule, is considerably less than the rate of spin-spin relaxation. Hence, in analyzing paramagnetic masers it is sufficient to limit ourselves to considering the range of intensities of the pumping radiation in which

$$\frac{1}{T_2^2} \gg |A_{13}|^2 \gg \frac{1}{T_2 T_1}. \tag{3.30}$$

Note that it is quite difficult to violate the left side of inequality (3.30) under real conditions for paramagnetic masers. Therefore condition (3.30) is not a limitation in practice. When condition (3.30) is satisfied, taking into account the definition of γ_{ik} and the solution of the rate equations by expansion in powers of the ratio T_2/T_1, we obtain from expression (3.27)

$$\varkappa_{23} = j \frac{I\Delta_{23}^0 T_2}{1 + \delta_{32}^2 T_2^2} \left[1 - j\delta_{32}T_2 + \frac{T_2}{T_1} \frac{\Delta_{13}^0}{I\Delta_{23}^0} \frac{(1 - j\delta_{31}T_2)(1 + \delta_{32}^2 T_2^2)}{1 - \delta_{21}\delta_{32}T_2^2 + j\delta_{31}T_2} - \frac{T_2}{T_1} S \frac{1 - j\delta_{32}T_2}{1 - \delta_{21}\delta_{32}T_2^2 + j\delta_{31}T_2} \left(1 - \frac{1}{1 + j\frac{\varkappa_{21}(1 - \delta_{21}\delta_{32}T_2^2 + j\delta_{31}T_2)}{I\Delta_{23}^0 T_2(1 + j\delta_{32}T_2)}} \right) \right], \tag{3.31}$$

where we have introduced the symbol S for the saturation factor $|A_{13}|^2 T_2 T_1$. It follows from the relation obtained that at the free frequency the effect of the intensity and the frequency of the pumping radiation and also of the field of the radiation is reduced by a factor of T_1/T_2. As $T_2/T_1 \to 0$ formula (3.31) essentially agrees with the results obtained by Butcher [36], who calculated the nondiagonal element of the density matrix ρ_{23} neglecting the presence of ρ_{21}.

We will consider the coefficient $\varkappa_{21}$ in more detail. In free space the moment M_{21} causes the appearance of a field $H_{21} = -4\pi M_{21}$. Then in accordance with the definitions (3.10) and (3.26)

$$\varkappa_{21} = -\hbar/4\pi N |\mu_{21}|. \tag{3.32}$$

For paramagnetic crystals with well resolved transitions $2 \to 1$, $\varkappa_{21} \approx T_2$ and consequently, for a careful analysis it is necessary to take into account the last term in the round brackets of expression (3.31). If the transition is forbidden the expression in the round brackets tends to unity.

In cavity-resonator and traveling-wave masers with slow-wave structures the operating material is situated in a multiple-frequency resonator. The field radiated by the moment M_{21} in certain types of oscillations with a Q-factor Q_i with a natural frequency ω_{0i}, is determined by the equation $\ddot{H}_{21} + \frac{\omega_{0i}}{Q_i}\dot{H}_{21} + \omega_{0i}^2 H_{21} = -4\pi \ddot{M}_{21}$. Assuming the presence of resonance types of oscillations at the pumping, signal, and free frequencies the coefficient $\varkappa_{21}$ can be represented in the form

$$\varkappa_{21} = -\frac{\hbar}{4\pi N |M_{21}|^2}\left(\frac{1}{jQ} - F\right), \tag{3.33}$$

where the effective Q-factor Q is determined by the Q-factors of the resonator Q_{21}, Q_{32}, and Q_{31} at the frequencies Ω_{21}, Ω_{32}, and Ω_{31}, respectively:

$$1/Q = 1/Q_{21} + \frac{\Omega_{32}}{\Omega_{21}} 1/Q_{32} + \frac{\Omega_{31}}{\Omega_{21}} 1/Q_{31},$$

and the factor $F = [2\Omega_{21}^2 - \Omega_{31}^2 - \Omega_{32}^2]/\Omega_{21}^2$. When $\Omega_{21} \gg \Omega_{32}$, $F \approx 1$. We will assume that $Q \gg 1$ and that $F = 1$. Then $\varkappa_{21} \approx \hbar/4\pi N|\mu_{21}|^2$. For well resolved transitions $\varkappa_{21}/I\Delta_{23}^0 T_2 \approx 1$, and the expression in the round brackets in formula (3.31) becomes essentially $(1 + j)/2$.

Then, taking into account in the terms of the order of T_2/T_1 only the frequency difference δ_{31} and neglecting its effect on the modulus of these correction terms, we obtain approximately

$$\varkappa_{23} = j\frac{I\Delta_{23}^0 T_2}{1+\delta_{32}^2 T_2^2}\left[1 - j\delta_{32}T_2 + \frac{T_2\Delta_{13}^0}{T_1 I\Delta_{23}^0}(1 - 2j\delta_{31}T_2) - \frac{1+j}{2}\frac{T_2}{T_1}S(1 - j\delta_{31}T_2)\right]. \tag{3.34}$$

If the free transition is forbidden the last term of this formula must be written in the form $(T_2/T_1)S(1 - j\delta_{31}T_2)$.

Under practical conditions, obviously, we always have $S \gg \Delta_{13}^0/I\Delta_{23}^0$, and we can write finally that in the case of a well resolved free transition

$$\frac{1}{Q} = -\eta\frac{|\mu_{23}|^2}{\hbar}N\frac{I\Delta_{23}^0 T_2}{1+\delta_{32}^2 T_2^2}\left[1 - j\delta_{32}T_2 - \frac{T_2}{T_1}S\frac{j+1-j\delta_{31}T_2}{2}\right]. \tag{3.35}$$

In the case when $|\mu_{21}|^2 = 0$

$$\frac{1}{Q_B} = -\eta\frac{|\mu_{23}|^2}{\hbar}N\frac{I\Delta_{23}^0 T_2}{1+\delta_{32}^2 T_2^2}\left[1 - j\delta_{32}T_2 - \frac{T_2}{T_1}S(1 - j\delta_{31}T_1)\right]. \tag{3.36}$$

We see therefore that the phase of the Q-factor of the active material is determined in particular by the pumping frequency. This small effect can probably be treated as a certain shift of the center of gravity of the distribution of the particles in the upper energy level common to the signal and pumping transitions, due to an inaccurately tuned pump, i.e., a pump which transfers the particles to this level but not to the exact center of this level. The spin-spin relaxation time is small and the particles are rapidly redistributed over the level, but in view of condition (3.30) this redistribution is unable to take place completely. This explains why for a suffiently deep saturation of the pumping transition the effect of the parameters of the pumping radiation on the phase of $1/Q_B$ is reduced by a factor T_1/T_2.

Before determining the parameters of masers with the help of the Q-factor Q_B we will consider in more detail the properties of this quantity. For practical applications the stabilities of masers are large. The fluctuations in the phase and in the modulus of $1/Q_B$ determine the phase and amplitude stability of masers. By taking into account the terms of the order of T_2/T_1 in the expression for $1/Q_B$ we can find the effect of the instability of the pumping radiation. In effect as $T_2/T_1 \to 0$ the quantity $1/Q_B$ can be represented in the form

$$1/Q_B = -1/Q_B^0 \frac{1 - 2j\frac{\Delta\nu}{\Delta\nu_l}}{1 + \left(2\frac{\Delta\nu}{\Delta\nu_l}\right)^2}, \tag{3.37}$$

where $\Delta\nu \equiv \nu - \nu_l$ is the frequency difference between the signal frequency ν and the line frequency $\nu_l \equiv \nu_{32}$, and for the modulus of the negative Q-factor of the active material at the center of the line of the signal transition we have introduced the notation

$$\left|\frac{1}{Q_B^0}\right| = \eta \frac{|\mu_{23}|^2}{\hbar} N I \Delta_{23}^0 T_2 \tag{3.38}$$

and we have taken into account that the spin-spin relaxation time T_2 is connected with the linewidth $\Delta\nu_l$ which has a Lorentz form, by the relation

$$T_2 = 1/\pi\Delta\nu_l. \tag{3.39}$$

The modulus of the Q-factor of the active material is

$$\left|\frac{1}{Q_B}\right| = \frac{1}{Q_B^0} \frac{1}{\sqrt{1 + \left(2\frac{\Delta\nu}{\Delta\nu_l}\right)^2}}. \tag{3.40}$$

For a frequency difference between the signal and the central frequency of $\Delta\nu = \pm \frac{1}{2}\sqrt{3}\,\Delta\nu_l$ the value of the modulus of $1/Q_B$ is reduced by a factor of 2. The phase of the Q-factor is determined by the relation

$$\tan\varphi = -2\Delta\nu/\Delta\nu_l. \tag{3.41}$$

When $\Delta\nu = \pm\frac{1}{2}\sqrt{3}\,\Delta\nu_l$ the phase shift is 60°. When the signal frequency is accurately tuned to the frequency of the spectral line the Q-factor of the active material is real and its phase is equal to zero. In the approximation considered $T_2/T_1 \to 0$, the phase instability $\delta\varphi$ is determined by the instability $\delta\nu_l$ of the frequency of the signal transition line ν_{32}. We see from expression (3.41) that in this case the small fluctuations in phase for a signal tuned to resonance equals

$$\delta\varphi = 2\frac{\delta\nu_l}{\Delta\nu_l}. \tag{3.42}$$

The instabilities in the frequency of the line $\delta\nu_l$ also cause the appearance of amplitude instabilities

$$\frac{\delta|Q_B|}{|Q_B|} = \frac{2\Delta\nu/\Delta\nu_l}{1 + \left(2\frac{\Delta\nu}{\Delta\nu_l}\right)^2} \frac{\delta\nu_l}{\Delta\nu_l}. \tag{3.43}$$

With accurate tuning these fluctuations disappear. We note at once that stabilization of the magnetic field, the use of permanent magnets, or use of magnets with superconducting windings, eliminates the problem of the phase and amplitude instabilities of paramagnetic masers due to instabilities in the frequencies of the signal transition line. Amplitude fluctuations can also be caused by instabilities in the quantity $1/Q_B^0$. As can be seen from formula (3.38) these instabilities are caused only by the instabilities of the inversion I and of the equilibrium difference in populations Δ_{23}. For strong saturation of the pumping transition the inversion is stable. This is one of the attractive properties of masers. Fluctuations in the equilibrium difference in the relative populations may be due to instability of the temperature of the operating material. When $h\nu \ll kT$ these fluctuations equal

$$\frac{\delta Q_B^0}{Q_B} = \frac{\delta T}{T}. \tag{3.44}$$

If the working material is placed in a bath of liquid helium, the temperature stability is determined by the stability of the pressure of evaporating helium and can be very large. By using special pressure-stabilizing devices one can obtain $\delta T/T \approx 10^{-4} - 10^{-5}$, for free evaporation of the helium into the atmosphere $\delta T/T$ is approximately 10^{-3}, and by using modern closed-cycle helium liquifiers one can obtain $\delta T/T = 10^{-2}$.

Therefore, the phase fluctuations due to fluctuations in the frequency of the absorption line relative to the signal frequency are completely eliminated by using magnets with superconducting windings. When condition (3.30) is satisfied the instabilities in the power of the pumping radiation do not lead to any noticeable fluctuations in the inversion. The temperature of the working material in principle can be stabilized to a high degree of accuracy. Therefore, in the approximations of [33, 36], the Q-factor of the active material can be assumed to be extremely stable. However, the more accurate expressions (3.35) and (3.36) enable us to determine the instability of the phase and amplitude of $1/Q_B$ due to changes in the frequency and intensity of the pumping radiation. Thus, for the most favorable case of a forbidden free transition we obtain from formula (3.36) that the change in phase due to a small change in the frequency of the pumping radiation $\delta\nu_p$ equals

$$\delta\varphi = \frac{T_2}{T_1} S \frac{\delta\nu_p}{\Delta\nu_l}. \tag{3.45}$$

For inaccurate tuning ($\delta_{31} \neq 0$) the fluctuations in S also lead to fluctuations in the phase of Q_B. In the case described by formula (3.35) fluctuations in S lead to fluctuations in the phase for accurate tuning also.

For an EPR line width of 50 MHz a detuning in the pumping frequency of 5 MHz for $S = 10$ leads, according to formula (3.45), to a change in phase of $10^{-5} - 10^{-7}$ radian, which agrees with the experimental facts of the high phase stability of masers.

Similarly, from formula (3.36) it is easy to take into account the instabilities of the modulus of $1/Q_B$, due to fluctuations in the pumping intensity even for saturation conditions. For the central frequency

$$\frac{\delta Q_B^0}{Q_B} = \frac{T_2}{T_1} \delta S. \tag{3.46}$$

When $S = 10$ even a 10% change in the power of the pumping radiation leads to $\delta Q_B^0/Q_B \approx 10^{-5} - 10^{-7}$, which is below the level of the fluctuations due to instabilities in the temperature of the operating material (3.44), and it can therefore be ignored.

Note that for saturation factors $S = 10-100$ we must nevertheless take into account the effect of pumping instabilities on the value of the inversion. When the pumping radiation is tuned to resonance formula (3.17) for the inverted population difference Δ_{32} can be represented in the form

$$I = \frac{I_0 S - 1}{S + 1} \approx I_0 - 2/S, \tag{3.47}$$

where $I_0 = T_2'/T_1$ [see formula (3.19)] is the maximum attainable inversion. From expressions (3.38) and (3.47) it follows that

$$\frac{\delta Q_B^0}{Q_B^0} = \frac{2}{I_0 S} \frac{\delta S}{S}. \tag{3.48}$$

Since the coefficient $2/I_0$ is of the order of unity, for $S = 10$ a 1% change in the pumping power leads to the value $\delta Q_B^0/Q_B^0 = 10^{-3}$.

Therefore, taking into account terms of the order of T_2/T_1 in the calculation of the susceptibility of the operating material has led to expressions which enable us to determine the effect of instabilities of the pumping frequency on the phase stability of the Q-factor of the operating material.

Estimates made on the basis of the formula obtained in this section lead to the conclusion that masers are very stable, which, together with their high sensitivity, is one of their chief advantages.

To conclude this chapter we will discuss the numerical estimates of the value of Q_B^0, and also the values of the pumping power P_p, required to obtain a given saturation factor S. We will consider the quantities in formula (3.38). We will assume that the filling is maximum, i.e., that $\eta = 1$. For the densities of paramagnetic ions in ruby-type crystals usually used $N = 2 \cdot 10^{19}$ cm^{-3} and $T_2 = 10^{-8}$ sec. Inversion is usually confined within the limits $I = 1 - 4$. We will assume that $I = 2$. If the signal transition is well resolved $|\mu_{32}|^2 \approx (g\beta)^2 \approx 4 \cdot 10^{-40}$ (erg/Oe)^2. Then $|\mu_{32}|^2/\hbar \approx 4 \cdot 10^{-13}$ cm$^3 \cdot$ sec^{-1}. In the centimeter band for three energy levels at a temperature $T = 4.2°K$ we can assume that $\Delta_{23}^0 \approx h\nu/3kT$. When $\nu = 10^{10}$, $\Delta_{23}^0 = 0.03$. As a result formula (3.38) gives a value of the modulus of the Q-factor of the active material of $Q_B^0 \approx 250$, which agrees well with the experimental data. The use of materials similar to rutile leads to $Q_B^0 \approx 50$.

To estimate the power of the source of pumping radiation we will relate the intensity of the high-frequency magnetic field of the pump H_p^2 in the pumping resonator to the loaded Q-factor Q_p and the pump power fed to the resonator input, by the relation

$$Q_p = \frac{2\pi \nu_p}{P_p} \frac{H_p^2}{8\pi} V_p, \tag{3.49}$$

where V_p is the volume of the resonator for the pumping radiation.

From formulas (3.10), (3.31), and (3.49) we obtain that

$$P_p = \frac{\hbar}{|\mu_{32}|^2} \frac{V_p \nu_p}{T_2 T_1 Q_p} S. \tag{3.50}$$

If $T_1 = 0.1$ sec, $Q_p = 10^3$, $V_p = 1$ cm^3, $\nu_p = 3 \cdot 10^{10}$ Hz, and $|\mu_{31}| = 0.1$ gβ then $P_p = S \cdot 10^{-3}$ W. Therefore, sufficiently deep saturation of the pumping transition (S = 10-100) can be achieved for fairly small pumping powers, which is very important for the successful development of masers.

CHAPTER IV

Microwave Masers

Although a medium with a negative absorption is necessary in order to construct an amplifier, it is not sufficient in itself. For the successful design of a maser it is necessary to ensure a certain interaction between the active material and the amplified radiation. Various methods of realizing stimulated emission by the active material are possible. Suppose the active material is placed in a waveguide. Then the electromagnetic wave propagating along the waveguide will increase exponentially. However, to obtain appreciable amplification requires filling a long length of the waveguide with active material. In practice it is very difficult to realize such a system. It is therefore necessary to increase considerably the degree of interaction between the active material and the amplified radiation. In one of these methods high Q-factor cavity resonators are used, in which the radiation interacts with the material under standing-wave conditions, traveling through the same specimen of material many times. In another method the radiation is propagated as a traveling-wave in a comparatively long specimen. However, in this case the specimen is placed in a slow-wave structure, in which the radiation is propagated with the group velocity, which is very much less than the speed of light. As a result of this, the interaction between the radiation and the active material becomes very large while the dimensions of the system as a whole remain reasonably small. In this chapter we will consider the properties of the cavity masers and traveling-wave masers obtained in this way [16, 38-42].

§1. Traveling-Wave Masers

As a rule, to reduce the dimensions in traveling-wave masers periodic slow-wave structures are used. For our purpose the consideration of the propagation of waves in periodic slow-wave structures is conveniently carried out by expressing the propagation constant in a periodic waveguide with losses in terms of the complex Q-factor of the spatial period of the slow-wave structure. The Q-factor of this period, as usual, is defined as the ratio of the energy stored in it to the complex power dissipated. In the theory of slow-wave structures it is well known that the amplitude attentuation factor of a wave propagating in such a waveguide with a group velocity u is given by the relation [43]

$$\alpha = \frac{2\pi \nu}{2Qu}. \tag{4.1}$$

This relation was generalized in [36] to the case of systems the real part of the Q-factor of which is negative. In this case the definition considered in the previous chapter of the Q-factor of the active material Q_B remains true. Then, taking into account the Q-factor Q_0, which describes the ohmic losses of energy, $1/Q = 1/Q_0 + 1/Q_B$, and the voltage transmission coefficient for a slow-wave structure of length l equals

$$K = e^{-\left(\frac{1}{Q_0} + \frac{1}{Q_B}\right)\frac{2\pi \nu l}{2u}}. \tag{4.2}$$

In the further analysis of the expression for Q_B (3.35) or (3.36) it is convenient to represent it in the form

$$1/Q_B = -\frac{1 - j\varphi}{Q_m}, \tag{4.3}$$

where

$$1/Q_m = \mathrm{Re}\,\frac{1}{Q_B} = \frac{1}{Q_B^0 (1 + \delta_{32}^2 T_2^2)}$$

and

$$\varphi = \delta_{32}T_2 - \frac{T_2}{T_1}S\delta_{31}T_2,$$

for the case when formula (3.36) is used or $\varphi = \delta_{32}T_2 + \frac{T_2}{T_1}S\frac{1-\delta_{31}T_2}{2}$ for the case when formula (3.35) is used. In accordance with the results of the discussion of formulas (3.35) and (3.36) in § 3, Chapter III, we will not take into account terms of the order of T_2/T_1 in the expression for the real part of $1/Q_B$.

From formulas (4.3) and (4.2) it follows that the voltage gain equals

$$K = \exp\left\{\left[-1/Q_0 + \frac{1-j\varphi}{Q_B^0(1+\delta_{32}^2 T_2^2)}\right]\frac{2\pi\nu l}{2u}\right\}. \tag{4.4}$$

Introducing the retardation of the group velocity $m = C/u$ and the number n of wavelengths in free space λ_0 in the length of the system $n = l/\lambda_0$, we can write equation (4.4) in the form

$$K = \exp\left\{mn\pi\left[\left(-1/Q_0 + 1/Q_B^0 \frac{1}{1+\frac{4(\nu-\nu_l)^2}{\Delta\nu_l^2}}\right) - j\frac{\varphi}{Q_B^0\left(1+\frac{4(\nu-\nu_l)^2}{\Delta\nu_l^2}\right)}\right]\right\}, \tag{4.5}$$

where ν_l denotes the central frequency of the line of the signal transition and we have taken into account the definitions (3.11) and (3.39) for δ_{32} and T_2. The first term of formula (4.5) describes the amplitude characteristic of the traveling-wave maser, and the second term describes the phase characteristic.

The power gain equals

$$G(\nu) = |K|^2 = \exp\left\{2\pi mn\left(\frac{1}{Q_B^0\left(1+\frac{4(\nu-\nu_l)^2}{\Delta\nu_l^2}\right)} - \frac{1}{Q_0}\right)\right\}. \tag{4.6}$$

This expression is conveniently expressed in the decibel scale

$$G_{dB}(\nu) = \frac{27.3}{Q_B^0}\frac{mn}{1+\frac{4(\nu-\nu_l)^2}{\Delta\nu_l^2}} - L_{dB}, \tag{4.7}$$

Where L_{dB} represents the ohmic losses in the slow-wave structure, expressed in decibels.

At resonance

$$G_{dB}(0) = \frac{27.3}{Q_B^0}mn - L_{dB}. \tag{4.8}$$

The phase characteristic of the traveling-wave maser is such that the phase advance of the output signal, as follows from (4.5) and (4.6), equals

$$\Phi = \frac{1}{2}\varphi \ln\frac{G(\nu)}{L}, \tag{4.9}$$

where the ohmic losses in the slow-wave structure $L = e^{-2\pi mn/Q_0}$. In the case of a strongly forbidden free transition

$$\varphi = 2\frac{\nu-\nu_l}{\Delta\nu_l} - 2\frac{T_2}{T_1}S\frac{\nu_p-\nu_p^0}{\Delta\nu_l}, \tag{4.10}$$

where ν_p is the frequency of the pumping radiation, and ν_p^0 is the central frequency of the pumping transition line. Consequently, in this case

$$\Phi = \frac{\nu - \nu_l}{\Delta \nu_l} \ln \frac{G(\nu)}{L} - \frac{T_2}{T_1} S \frac{\nu_p - \nu_p^0}{\Delta \nu_l} \ln \frac{G(\nu)}{L}. \qquad (4.11)$$

Therefore, the frequency and phase characteristics of traveling-wave masers are determined by the attainable gain.

It follows from (4.7) that the transmission bandwidth of a traveling-wave maser between the 3 dB levels equals

$$\Delta \nu = \Delta \nu_l \sqrt{\frac{3}{G_{dB} + L_{dB} - 3}}. \qquad (4.12)$$

For slow-wave structures consisting of posts the value of m can reach 100-200 for ohmic losses of the order of 10 dB. Then for $Q_B^0 \approx 250$ values of n = 2-3 are sufficient to obtain a gain of 20 dB. In this case (G = 20 dB, L = 10 dB) the transmission band of the maser in accordance with formula (4.12) equals $\Delta \nu = 0.30 \Delta \nu_l$.

Note that the formulas given for the gain and transmission bandwidth of traveling-wave masers are only true when a traveling-wave in fact exists in the amplifier. For this the maser must possess nonreciprocal, unidirectional amplification, in which only the wave traveling from the input to the output is amplified. To suppress unknown reflections the amplifier must exhibit nonreciprocal absorption of the backward wave traveling from the output to the input. If we assume that the slow-wave structure is uniform and that reflection only occurs at the beginning and end of the structure, the value of the attenuation of the backward wave L_{back} must satisfy the condition

$$L_{back} \gg G \Gamma_{in}^2 \Gamma_{out}^2, \qquad (4.13)$$

where Γ_{in} and Γ_{out} are the reflection coefficients of the input and output. If reflections from the input and output are not suppressed standing waves are set up in the system, positive backward waves occur, and the traveling-wave maser becomes a cavity maser (see § 2, Chapter IV).

The absence in traveling-wave masers of positive feedback is responsible for the relatively high stability of these masers.

Considering amplification at the central frequency we obtain from (4.7) that the relative instabilities are given by the expression

$$\frac{\delta G(0)}{G(0)} = \frac{\delta Q_B^0}{Q_B^0} \ln \frac{G}{L}. \qquad (4.14)$$

Similarly, the fluctuations in phase are given by

$$\delta \Phi = \frac{1}{2} \delta \varphi \ln \frac{G}{L}. \qquad (4.15)$$

The quantities $\delta Q_B^0/Q_B^0$ and $\delta \varphi$ are considered in detail in Chapter III.

When G/L equals 10^3 (G_{dB} = 20 dB, L_{dB} = 10 dB) the estimates in § 3, Chapter III, for these quantities lead to instabilities of the maser of $\delta G_0/G_0 \approx 7 \cdot 10^{-3}$, $\delta \phi \approx 7(10^{-5} - 10^{-7})$.

We will consider the noise of traveling-wave masers. In § 3, Chapter I, we presented formulas for the spectral density of the noise of the spontaneous radiation. The thermal radiation of the walls of the waveguide of the maser is added to the noise of the spontaneous radiation. Then the equation for the spectral density of the energy flow in the traveling-wave system can be written in the form

$$\frac{dS}{dZ} = -(\alpha_1 - \alpha_2) S + \alpha_1 P_{\nu 0} + \alpha_2 P_{\nu sp}, \qquad (4.16)$$

where Z is a coordinate measured along the direction of propagation of the traveling-wave, α_1 is the attenuation constant in the waveguide walls, α_2 is a constant of amplification, $P_{\nu 0}$ is the spectral density of the thermal (equilibrium) noise, and $P_{\nu\text{sp}}$ is given by formula (1.26). From formula (4.16) it is easy to obtain the spectral density of the inherent noise of the amplifier referred to its input

$$P_{\nu \text{in}} = \frac{G-1}{G} \frac{\alpha_2 P_{\nu\text{sp}} + \alpha_1 P_{\nu 0}}{\alpha_2 - \alpha_1}, \tag{4.17}$$

where $G = \exp\{(\alpha_2 - \alpha_1) l\}$, and l is the length of the maser. This relation can conveniently be written in another form, by expressing the parameters α_1 and α_2 in terms of the gain G_{dB} and the ohmic losses L_{dB}. Since $\alpha_1/(\alpha_2 - \alpha_1) = L_{\text{dB}}/G_{\text{dB}}$, we have

$$P_{\nu \text{in}} = \frac{G-1}{G} \left[P_{\nu\text{sp}} + (P_{\nu\text{sp}} + P_{\nu 0}) \frac{L_{\text{dB}}}{G_{\text{dB}}} \right]. \tag{4.18}$$

In the radio band and for large gains the effective temperature of the input noise of a traveling-wave maser is approximately equal to

$$T_{\text{in}} = |T_{\text{eff}}| + T_0 \frac{L_{\text{dB}}}{G_{\text{dB}}}. \tag{4.19}$$

At liquid-helium temperatures the second term in this formula can be neglected. Therefore, the inherent noise of traveling-wave masers is in fact very small and corresponds to an effective temperature of the order of 2-10°K. Note that as $L_{\text{dB}} \to 0$ formula (4.18) agrees with (1.3), which gives the minimum possible value of the spectral density of the input noise of a coherent amplifier. The properties of traveling-wave masers will be considered at the end of this chapter when they are compared with the properties of cavity masers.

§ 2. The Single-Resonator Maser

Suppose the active material is placed in a resonator which has a single coupling hole with supply feeders. The equivalent LC circuit is shown in Fig. 1. The inherent losses of the resonator are represented by the resistance R_0, and the negative losses introduced by the active material are represented by the resistance R_B. The wave impedance of the feeder line equals Z_0, and the coupling loop of the circuit with the line has a coefficient of mutual inductance M. Since the reflected signal is the amplified signal the gain is equal to the reflection coefficient. As is well known from the theory of long lines, the voltage reflection coefficient Γ of a terminating load with an impedance Z is given by the relation

$$\Gamma = \frac{\beta - 1}{\beta + 1}, \tag{4.20}$$

where $\beta = Z/Z_0$. In this case the power gain $G = |\Gamma|^2$. The value of Z is determined by the resistance increase when the load is inserted. It is easy to see from Fig. 1 that

$$Z = j\omega L_{\text{CB}} + \frac{\omega^2 M^2}{R_0 + R_B + j\omega L(1 - \omega_0^2/\omega^2)}, \tag{4.21}$$

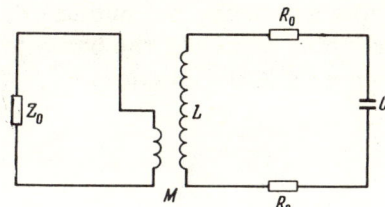

Fig. 1. Equivalent circuit of a single-resonator maser.

where $j\omega L_{CB}$ is the inductive reactance of the coupling loop and $\omega_0 = 1/\sqrt{LC}$ is the natural frequency of the circuit. Relating the Q-factor of the loaded resonator Q_0, of the active material Q_B, and of the coupling Q_{CB} with the respective values R_0, R_B, and the coupling resistance introduced into the circuit $\omega^2 M^2/Z_0$ by the relations

$$Q_0 = \frac{\omega_0 L}{R_0}, \quad Q_B = \frac{\omega_0 L}{R_B}, \quad Q_{CB} = \frac{\omega_0 L}{\omega^2 M^2/Z_0} \tag{4.22}$$

respectively, and neglecting the term $\omega L_{CB}/Z_0$, we obtain

$$\beta = \frac{1/Q_{CB}}{1/Q_0 + 1/Q_B + j\dfrac{\omega}{\omega_0}(1 - \omega_0^2/\omega^2)}. \tag{4.23}$$

In the further analysis it is convenient to introduce the relative detuning of the signal frequency ν with respect to the natural frequency of the resonator

$$X = 2\frac{\nu - \nu_0}{\nu_0}. \tag{4.24}$$

For frequencies which do not differ very greatly from the resonance frequency we can assume that $(\omega/\omega_0)(1 - \omega_0^2/\omega^2) \approx X$. Then, taking into account formula (4.3) for $1/Q_B$, we obtain

$$\beta = \frac{1/Q_{CB}}{1/Q_0 - 1/Q_m + i(X + \varphi/Q_m)}. \tag{4.25}$$

Introducing the relative difference between the signal frequency and the frequency of the signal transition line $\nu_l X_l = 2(\nu - \nu_l)/\nu_l$ and the Q-factor of the signal transition line $Q_l = \nu_l/\Delta\nu_l$, assuming the resonator to be tuned to the frequency of the line ($X = X_l$) and neglecting terms of the order of T_2/T_1, we obtain that the power gain of a reflection resonator maser equals

$$G(x) = \frac{[1/Q_0 - 1/Q_{CB} - 1/Q_B^0(1 + X^2 Q_l^2)]^2 + X^2[1 + Q_l/Q_B^0(1 + X^2 Q_l^2)]^2}{[1/Q_0 + 1/Q_{CB} - 1/Q_B(1 + X^2 Q_l^2)]^2 + X^2[1 + Q_l/Q_B^0(1 + X^2 Q_l^2)]^2}. \tag{4.26}$$

At resonance

$$G(0) = \frac{(-1/Q_0 + 1/Q_{CB} + 1/Q_B^0)^2}{(1/Q_0 + 1/Q_{CB} - 1/Q_B^0)^2}. \tag{4.27}$$

The gain is determined, in particular, by the degree of coupling of the maser resonator with the line. In what follows it will be useful to have formula (4.28) which enable us to determine the value of the coupling Q-factor as a function of the required gain:

$$1/Q_{CB} = (1/Q_B^0 - 1/Q_0)\frac{G^{1/2}(0) + 1}{G^{1/2}(0) - 1}. \tag{4.28}$$

The frequency characteristic of the maser is determined by formula (4.20). Its shape is close to the frequency characteristic of a single-frequency tuned circuit, although it is also somewhat more complex. The bandwidth can be estimated approximately by equating to one another the terms in the denominator of (4.26). For frequencies which are not outside the limits of the line width, i.e., for $X_l^2 Q_l^2 \ll 1$, we can obtain a relation which connects the bandwidth of the maser at the level $G = \frac{1}{2}G(0)$ with the value of the resonance gain [33]:

$$(G^{1/2}(0) - 1)\Delta\nu = \frac{2\nu}{Q_l}\frac{1/Q_B^0 - 1/Q_0}{1/Q_B^0 + 1/Q_l}. \tag{4.29}$$

Note that for a transmission maser in the optimum case of the same coupling Q-factor for the input and output coupling holes the value of $(G^{1/2}(0)-1)\Delta\nu$ is half of this. Relation (4.29) is approximate. For the case when the spectral line is much wider than the bandwidth of the loaded resonator we can obtain rigorously [16]

$$(G^{1/2}(0)-1)\Delta\nu = 2\nu_0(1/Q_B^0 - 1/Q_0). \tag{4.30}$$

When $Q_l \ll Q_B^0$ expression (4.29) transforms into (4.30). The gain is large when $1/Q_B^0 \lesssim 1/Q_0 + 1/Q_{CB} = 1/Q_H$. Consequently, the correction for the line width can be neglected when $Q_l \ll Q_H$. In this case the bandwidth of the maser is determined by the bandwidth of the loaded resonator

$$\Delta\nu \approx 2 \frac{\Delta\nu_H}{G^{1/2}(0)}. \tag{4.31}$$

In the converse case, i.e., when $Q_l \gg Q_H \approx Q_B^0$, we obtain that the bandwidth of the maser is determined by the width of the regenerated line of the resonance transition

$$\Delta\nu \approx 2 \frac{\Delta\nu_l}{G^{1/2}(0)}. \tag{4.32}$$

This formula gives the maximum value of the product $G^{1/2}(0)\Delta\nu$ for a single-resonator reflection maser. In this case we must satisfy the condition $Q_B^0 \ll Q_l = \nu/\Delta\nu_l$. Since the spin-spin relaxation time T_2, which determines the line width, is practically independent of the frequency, other conditions being equal at high frequencies the limiting value of $G^{1/2}(0)\Delta\nu$ for the given material is rapidly reached. In practice it is found that formula (4.31) corresponds to the decimeter band and formula (4.32) corresponds to the millimeter band. For a gain of 20 dB in this case $\Delta\nu = 0.2\Delta\nu_l$. Under practical conditions usually formula (4.32) is represented by $G^{1/2}(0)\Delta\nu \approx$ (50-100) MHz, and formula (4.31) is represented by $G^{1/2}(0)\Delta\nu \approx \nu_0/(200\text{-}300)$ MHz.

The presence of positive feedback in resonator masers leads to a stronger dependence of the bandwidth on the gain than in traveling-wave masers.

The phase characteristic of a reflection resonator maser is determined by the phase of the reflection coefficient Γ. When $Q_0 \gg Q_{CB}$

$$\Gamma = \frac{1/Q_{CB} + 1/Q_m - i(X + \varphi/Q_m)}{1/Q_{CB} - 1/Q_m + i(X + \varphi/Q_m)}. \tag{4.33}$$

For large amplifications we can assume approximately that the phase is determined by the denominator of expression (4.33). We then obtain for the phase of the amplified signal the relation

$$\tan\Phi = -\frac{1}{2} Q_{CB}(G^{1/2} + 1)(X + \varphi/Q_m). \tag{4.34}$$

In the particular case already considered of a strongly forbidden free transition the phase of the Q-factor of the material is given by formula (4.10). Then

$$\tan\Phi = \frac{\nu - \nu_0}{\Delta\nu_H}(G^{1/2} + 1) + \frac{\nu - \nu_l}{\Delta\nu_l}(G^{1/2} - 1) - \frac{T_2}{T_1} S \frac{\nu_p - \nu_p^0}{\Delta\nu_l}(G^{1/2} - 1), \tag{4.35}$$

where we have taken into account that $Q_{CB} \approx Q_H = \nu/\Delta\nu_H$. The phase shift is proportional to the voltage amplification and occurs when the signal is detuned relative to the central frequency of the resonator ν_0, or the line frequency ν_l (of the order of T_2/T_1) when the pumping radiation is detuned relative to the frequency ν_p^0. The relation obtained from the corresponding expression (4.11) for a traveling-wave maser is distinguished by a stronger dependence on the gain and by the presence of an additive term proportional to the detuning of the signal frequency from the resonator frequency.

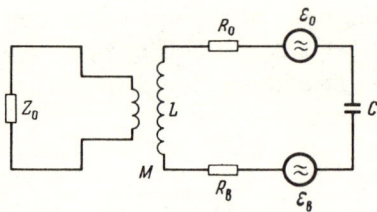

Fig. 2. Fluctuation emf's in the equivalent circuit of a single-resonator maser.

The higher degree of regeneration of resonator masers leads to an increase in the instability in comparison with traveling-wave masers for the same stability of the parameters of the Q-factor of the material. In addition, for resonator masers we must take into account the change in the coupling Q-factor Q_{CB}. For amplification at the central frequency the small relative fluctuations $\delta G(0)/G(0)$ are given by the expression [for fairly large $G(0)$]

$$\frac{\delta G(0)}{G(0)} = G^{1/2}(0)\left[\frac{\delta Q_{CB}}{Q_{CB}} + \frac{\delta Q_B^0}{Q_B^0}\right]. \tag{4.36}$$

Similarly, small fluctuations in phase are given by

$$\delta\Phi = G^{1/2}(0)\frac{\delta\nu}{\Delta\nu_H} + \frac{1}{2}G^{1/2}(0)\,\delta\varphi. \tag{4.37}$$

Comparing (4.37) and (4.36) with expression (4.14) and (4.15) we see that the dependence of the instabilities in the resonator maser on the gain is qualitatively different. However, this leads to considerable differences only for large gains.

It is convenient to determine the noise of a single-resonator maser of the reflection type from its equivalent circuit, in which the fluctuation emf's are taken into account (Fig. 2). The generator of emf ε_0 represents the thermal radiation of the resonator walls, and the emf ε_B corresponds to spontaneous noise. The values of the spectral densities of the squares of the voltages ε_0 and ε_B are given by the formulas (1.24) and (1.25), respectively. In this case, in (1.24) we must take $R = R_0$ and $T = T_0$, and in (1.25) $|R| = R_B$, where the indexes 0 and B refer to the resonator and the material respectively. The noise generated in the resonator is amplified and is fed into the feeder with a wave impedance of Z_0 via the coupling transformer with a coefficient of mutual inductance M. The power of the output noise is equal to the noise power dissipated in the resistance of Z_0. At the resonance frequency Kirchhoff's equations for the circuit of Fig. 2 can be written in the form

$$\begin{aligned} 0 &= Z_0 I_0 + j\omega M I_p, \\ \varepsilon_0 + \varepsilon_B &= I_p(R_0 - R_B) + j\omega M I_0, \end{aligned} \tag{4.38}$$

where I_0 is the current through the wave impedance Z_0, and I_p is the current in the circuit. Hence, taking into account the definitions of the Q-factors Q_0, Q_{CB}, and Q_B^0

$$I_0 = -j\frac{\omega M}{Z_0}\frac{1}{\omega L}\frac{\varepsilon_0 + \varepsilon_B}{1/Q_0 + 1/Q_{CB} - 1/Q_B^0}. \tag{4.39}$$

The spectral density of the power dissipated by the current I_0 in the impedance Z_0 equals $\overline{Z_0 I_{0\nu}^2}$. Hence, it is easy to obtain that taking into account (1.24) and (1.25) the spectral density of the output noise equals

$$P_{\nu\,\text{out}} = \left(\frac{2/Q_{CB}}{1/Q_0 + 1/Q_{CB} - 1/Q_B^0}\right)^2\left(\frac{Q_{CB}}{Q_0}\frac{h\nu}{e^{h\nu/kT_0}-1} + \frac{Q_{CB}}{Q_B^0}\frac{h\nu e^{h\nu/k|T_{\text{eff}}|}}{e^{h\nu/k|T_{\text{eff}}|}-1}\right). \tag{4.40}$$

In practice, the case when $Q_0 \gg Q_B^0$ and $Q_{CB} = \frac{G^{1/2}(0)-1}{G^{1/2}(0)+1}Q_B^0$ is important. Then referred to the input of the maser we obtain

$$P_{\nu\text{in}} = \frac{G(0)-1}{G(0)} \left(\frac{Q_B^0}{Q_0} \frac{h\nu}{e^{h\nu/kT_0}-1} + \frac{h\nu e^{h\nu/k|T_{\text{eff}}|}}{e^{h\nu/k|T_{\text{eff}}|}-1} \right). \tag{4.41}$$

Consequently, for large amplifications and for $h\nu \ll kT_0$, $k|T_{\text{eff}}|$ the effective temperature of the input noise of single-resonator masers equals

$$T_{\text{in}} = |T_{\text{eff}}| + \frac{Q_B^0}{Q_0} T_0, \tag{4.42}$$

which essentially agrees with the noise temperature of the traveling-wave maser [see formula (4.19)].

The previous discussion related to resonator masers in which, as a rule, resonator dimensions comparable with the wavelength of the excited wave are used. In such resonators the active material is concentrated mainly at the antinodes of the standing-wave field of the amplified radiation. However, the dimensions of the resonators can considerably exceed the wavelengths, and the active material used in them can have the form of a long rod. In such systems the amplified radiation is propagated in the active material in the form of a series of growing traveling waves. Essentially, such resonator masers are regenerative traveling-wave masers, the positive feedback in which is realized when there is reflection from the boundaries which separate the resonators from the feeders (or from free space). Hence, it is of interest to obtain expressions for the gain and the bandwidth of such amplifiers as a function of the power gain under traveling-wave conditions for one passage G_1, which characterizes the active material, and the power reflection coefficient at the boundary R, which characterizes the resonator. The results obtained are true not only for regenerative traveling-wave masers in the radio band but also for optical resonator masers with single-axial oscillation.

The discussion will be carried out assuming that the width of the signal transition line is fairly large in comparison with the bandwidth of the resonator. We will consider a transmission maser with a disc resonator [44] (a resonator of the Fabry-Perot interferometer type), a unidirectional transmission maser [45], in which the radiation is propagated round a circular path, and a reflection maser with a circulator. The relations which characterize these amplifiers will be obtained not by analyzing their equivalent circuits with lumped constants but by adding (taking the phases into account) the amplitudes of the fields which emerge from the resonators after multiple reflections from the mirrors which bound the resonators.

Summing the amplitudes of the beams which leave the Fabry-Perot resonator and assuming $G_1 R < 1$, we obtain that the complex voltage gain equals

$$K = (1-R)\sqrt{G_1} \frac{e^{j\omega'/c}}{1 - G_1 R e^{-j2\omega l/c}}. \tag{4.43}$$

Here we have assumed that the input and output mirrors are the same, and l denotes the length of the maser.

Consequently, the power gain is

$$G(\lambda) = \frac{(1-R)^2 G_1}{1 - 2RG_1 \cos 4\pi/\lambda + R^2 G_1^2}. \tag{4.44}$$

At resonance, i.e., when $\cos 4\pi l/\lambda = 1$,

$$G_0 = \frac{(1-R)^2 G_1}{(1-RG_1)^2}. \tag{4.45}$$

When R = 0 traveling-wave operation occurs and $G = G_1$.

When R = 1 the input radiation does not travel through the maser and G = 0. When R = $1/G_1$ generation occurs. The bandwidth of the maser can be estimated in the following way.

For a detuning from resonance by an amount $\Delta\nu/2$, $\cos 4\pi l/\lambda$ can be represented in the form

$$\cos 4\pi \frac{l}{\lambda_0}\left(1 + \frac{\Delta\nu}{2\nu}\right) = 1 - \frac{1}{2}\left(4\pi \frac{l}{\lambda_0} \frac{\Delta\nu}{2\nu}\right)^2.$$

Then the passband between the 3 dB levels is given by the formula

$$\frac{\Delta\nu}{\nu} = \frac{\nu}{2\pi l} \frac{1 - G_1}{\sqrt{G_1 R}}. \tag{4.46}$$

Expressing R in terms of $G(\lambda_0)$ and G_1 we obtain

$$\Delta\nu \sqrt{(G_0^{1/2} - G_1^{1/2})(G_1^{1/2} G_0^{1/2} - 1)} = \nu \frac{\lambda}{2\pi l} \frac{G_1 - 1}{G_1^{1/4}}. \tag{4.47}$$

Usually in the radio band $l \approx \lambda$ and G_1 differs very little from unity. Then the left side of this formula reduces to the product $\Delta\nu(G^{1/2} - 1)$, which was considered above [see formulas (4.29) and (4.30)]. In the case of practical interest when $G_0^{1/2} \gg G_1^{1/2}$, (4.47) reduces to

$$\Delta\nu G_0^{1/2} = \nu \frac{\lambda}{2\pi l} \frac{G_1 - 1}{G_1^{1/2}}. \tag{4.48}$$

When the amplification in one passage is increased the bandwidth of the maser is increased. This is explained by the fact that to attain the given G_0 for large G_1 less regeneration is necessary. In view of this the relative instabilities in the gain for a given G_0 are reduced as G_1 increases. We can obtain from (4.45) that for large G_0

$$\frac{\delta G_0}{G_0} = 2 \frac{G_1^{1/2}}{G_1 - 1} G_0^{1/2}\left(\frac{\delta G_1}{G_1} + \frac{1}{G_1^{1/2}} \frac{\delta R}{R}\right). \tag{4.49}$$

The connection between this formula and formula (4.36) is obvious.

The degree of regeneration can be reduced when approximating to traveling-wave conditions. If the amplified radiation propagates round a circular path, when the signal travels through the active material in only one direction the regeneration can be reduced. Such a closed optical path when using a straight ruby rod in a three-mirror system has been realized in the investigation of lasers [46, 47].

Summing the amplitude of the fields at the output of the system assuming $\sqrt{RG_1} < 1$, we obtain that

$$K = \frac{-\sqrt{R} + \sqrt{G_1}\, e^{-j\omega l/c}}{1 + \sqrt{RG_1}\, e^{-j\omega l/c}}, \tag{4.50}$$

where R is the reflection coefficient of the input mirror of the system. Consequently,

$$G(\lambda) = \frac{R - 2\sqrt{RG_1}\cos 2\pi l/\lambda + G_1}{1 - 2\sqrt{RG_1}\cos 2\pi l/\lambda + RG_1}. \tag{4.51}$$

At resonance, i.e., when $\cos 2\pi l/\lambda = 1$,

$$G_0 = \left(\frac{\sqrt{R} - \sqrt{G_1}}{1 - \sqrt{RG_1}}\right)^2. \tag{4.52}$$

In this case

$$\Delta\nu G^{1/2} = \nu \frac{\lambda}{\pi l} \frac{G_1 - 1}{G_1^{1/2}} \tag{4.53}$$

and
$$\frac{\delta G_0}{G_0} = \frac{G_1^{1/2}}{G_1 - 1} G_0^{1/2} \left(\frac{\delta G_1}{G_1} + \frac{1}{G_1^{1/2}} \frac{\delta R}{R} \right). \tag{4.54}$$

From a comparison of (4.54) and (4.53) with expressions (4.49) and (4.48) we see that in a unidirectional "ring" maser regeneration is actually weaker than in a transmission maser with a Fabry-Perot resonator.

The reflection regenerative traveling-wave maser which is bounded on one side by a completely reflecting mirror is very promising. On the side opposite the completely reflecting mirror is placed a partially transparent mirror which is simultaneously the input and output mirror of the maser.

Summing the amplitudes of the fields at the output of the system assuming $G_1\sqrt{R} < 1$ we obtain that

$$K = \frac{G_1 e^{-j\omega 2l/c} - \sqrt{R}}{1 - G_1 \sqrt{R} e^{-j\omega 2l/c}}, \tag{4.55}$$

where R is the reflection coefficient of the input mirror of the system. Consequently,

$$G(\lambda) = \frac{R - 2G_1 \sqrt{R} \cos 4\pi l/\lambda + G_1^2}{1 - 2G_1 \sqrt{R} \cos 4\pi l/\lambda + G_1^2 R}. \tag{4.56}$$

At resonance, i.e., when $\cos 4\pi l/\lambda = 1$,

$$G_0 = \left(\frac{G_1 - \sqrt{R}}{1 - G_1 \sqrt{R}} \right)^2. \tag{4.57}$$

Note that these formulas for the reflection maser are similar to the corresponding formulas for a ring maser in which G is replaced by $\sqrt{G_1}$ and the resonance length is changed.

When R = 1 there is no amplification in the reflection maser (G = 1) and when R = 0 there is a traveling wave traveling through the active material twice ($G = G_1^2$). Generation begins when $R = 1/G_1^2$. The value of the reflection coefficient required to attain a required gain for a given G_1 is given by the formula

$$R = \left(\frac{\sqrt{G_0} - G_1}{G_1 \sqrt{G_0} - 1} \right)^2. \tag{4.58}$$

To obtain $G_0 = 100$ for $G_1 = 3$ it is necessary to have R = 6%. Better use of the active material is achieved in the reflection maser. As a result of this, other things being equal, its bandwidth and stability are considerably better than those of the masers considered above. For large G_0

$$\Delta \nu G_0^{1/2} = \nu \frac{\lambda}{2\pi l} \frac{G_1^2 - 1}{G_1}, \tag{4.59}$$

$$\frac{\delta G_0}{G_0} = 2 \frac{G_1}{G_1^2 - 1} G_0^{1/2} \left(\frac{\delta G_1}{G_1} + \frac{1}{G_1} \frac{\delta R}{R} \right). \tag{4.60}$$

Therefore, the reflection maser is more broadband and correspondingly more stable than the transmission maser by a factor of $(G_1 + 1)/\sqrt{G_1}$. For $G_1 = 3$ this amounts to a factor of 2.34, and for $G_1 \approx 1$ the parameters of the reflection maser exceed the parameters of the transmission maser by a factor of 2, as already noted when discussing formula (4.29). In concluding this section we note that the main drawback of single-resonator masers is the fact that they are narrowband. Later, we will compare these types of masers with other forms of masers.

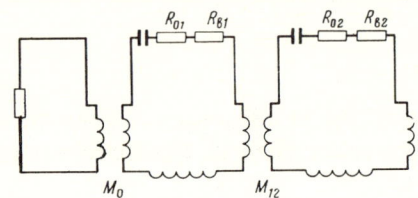

Fig. 3. Equivalent circuit of a two-resonator maser.

§3. Two-Resonator Masers

The bandwidth of resonator masers can be increased for a given gain by using two resonators coupled to one another [48]. We will consider two-resonator reflection masers. The equivalent circuit of Fig. 3 shows a two-resonator maser with series-connected resonators. From this circuit we can obtain the value of the impedance of the coupled resonators converted into the line

$$Z = \cfrac{\omega^2 M_0^2}{R_{01} + R_{B1} + j\omega L_1\left(1 - \cfrac{\omega_{01}^2}{\omega^2}\right) + \cfrac{\omega^2 M_{12}^2}{R_{02} + R_{B2} + j\omega L_2\left(1 - \cfrac{\omega_{02}^2}{\omega^2}\right)}}, \qquad (4.61)$$

where we have retained the notation of the previous section, and the indexes 1 and 2 relate to the first and second resonators respectively. Introducing the coupling coefficient between the resonators $\varkappa = \sqrt{M_{12}^2/L_1 L_2}$, and also the coupling Q-factor of the loaded circuits and the active material we can write the parameter β in the form

$$\beta = \cfrac{1/Q_{\text{CB}}}{1/Q_{01} - 1/Q_{m1} + i\left(X_1 + \cfrac{Q_l}{Q_{m1}} X_l\right) + \cfrac{\varkappa^2}{1/Q_{02} - 1/Q_{m2} + i\left(X_2 + \cfrac{Q_l}{Q_{m2}} X_l\right)}}. \qquad (4.62)$$

Here we have neglected the effect of terms of the order of T_2/T_1 on the form of the frequency characteristic of the maser. The quantity β enables us to determine the gain and the form of the frequency characteristic of the masers considered.

For example, in the particular case of a two-resonator maser with an input passive resonator, i.e., with an input resonator not containing any active material ($Q_{m1} = \infty$), for the same tuning of the resonators ($X_1 = X_2 = X_l$) and a value of the coupling coefficient $\varkappa = 1/Q_B^0$ we obtain that it is not the product $(G_0^{1/2} - 1)\Delta\nu$ that is constant, but the quantity $\Delta\nu \sqrt{G^{1/2}(0) - 1}$, which equals (for $Q_0 \gg Q_B^0$)

$$\Delta\nu \sqrt{G^{1/2}(0) - 1} = \frac{\nu_0}{Q_B^0 + Q_l} \sqrt{2(\sqrt{2} - 1)}. \qquad (4.63)$$

Comparison of (4.63) with expression (4.29) shows that a two-resonator maser for a given gain has a bandwidth approximately $\sqrt[4]{G(0)}$ times wider than a single-resonator maser.

If in the two-resonator maser the active material is in both resonators, for the same tuning ($X_1 = X_2 = X_l = X$) and assuming that $Q_{01} \approx Q_{02} \gg Q_{B1}^0 = Q_{B2}^0 = Q_B^0$, we obtain the following expression for the frequency characteristics of such a maser:

$$G(x) = \frac{\left[\frac{1}{Q_m Q_{\text{CB}}} + \varkappa^2 + \frac{1}{Q_m^2} - \left(1 + \frac{Q_l}{Q_m}\right)^2 X^2\right]^2 + \left(1 + \frac{Q_l}{Q_m}\right)^2 X^2 \left(\frac{1}{Q_{\text{CB}}} + \frac{2}{Q_m}\right)^2}{\left[\frac{1}{Q_m Q_{\text{CB}}} - \varkappa^2 - \frac{1}{Q_m^2} + \left(1 + \frac{Q_l}{Q_m}\right)^2 X^2\right]^2 + \left(1 + \frac{Q_l}{Q_m}\right)^2 X^2 \left(\frac{1}{Q_{\text{CB}}} - \frac{2}{Q_m}\right)^2}. \qquad (4.64)$$

At the central frequency, i.e., for X = 0,

$$G(0) = \left(\frac{1/Q_B^0 Q_{CB} + \varkappa^2 + 1/Q_B^{02}}{1/Q_B^0 Q_{CB} - \varkappa^2 - 1/Q_B^{02}}\right)^2. \qquad (4.65)$$

For a value of the coupling coefficient $\varkappa = 1/Q_B^0$ the value of $\Delta \nu$ is given by the formula

$$\Delta \nu \sqrt{G^{1/2}(0) - 1} = \frac{\nu}{Q_l + Q_B^0} 2\sqrt{\sqrt{2} - 1}. \qquad (4.66)$$

Formulas (4.63) and (4.66) differ from the results obtained in [49] by taking into account the finiteness of the width of the paramagnetic resonance line, which is essential in the centimeter and also in the decimeter waveband for high inversion (small Q_B^0).

Therefore, for a maser with two active resonators the bandwidth is $\sqrt{2}$ times wider than for a two-resonator maser with one active resonator. In addition, a maser with two active resonators is more stable. In fact, from (4.65) assuming that the Q-factors of the same kind of active material vary in the same way, we can obtain

$$\frac{\delta G(0)}{G(0)} = G^{1/2}(0)\left\{\frac{1 - \varkappa^2 Q_B^{02}}{1 + \varkappa^2 Q_B^{02}} \frac{\delta Q_B^0}{Q_B^0} + \frac{\delta Q_{CB}}{Q_{CB}}\right\}. \qquad (4.67)$$

This formula shows the advisability of choosing the coupling coefficient $\varkappa = 1/Q_B^0$ to eliminate the effect of fluctuations in the Q-factor of the active material. Complete compensation of fluctuations in gain, of course, cannot be obtained. The presence of the second resonator, naturally, does not reduce the effect of fluctuations in the coupling Q-factor.

The large amplitude stability of two-resonator masers was pointed out in [48]. However, the phase characteristics of two-resonator masers with two active resonators are also stable. From (4.62) we can obtain that the fluctuations in the phase of the gain due to instabilities in the phase of the Q-factor of the active material equal

$$\delta \Phi = \frac{1}{2} G^{1/2}(0) \frac{1 - \varkappa^2 Q_B^{02}}{1 + \varkappa^2 Q_B^{02}} \delta \varphi. \qquad (4.68)$$

The appearance of the factor $1 - \varkappa^2 Q_B^{02}$ in these formulas is due to the transformation of the impedances in the coupling transformer between the circuits M_{12} [see Fig. 3 and formula (4.61)]. Therefore, the instabilities in the resistances R_{B1} and R_{B2} are mutually compensated provided they are the same. Hence, it turns out that the instabilities of the two-resonator maser with an input passive resonator are the same as those for a single-resonator maser which also verifies the direct calculation.

The noise of a two-resonator maser can easily be found as in the case of a single-resonator maser from its equivalent circuit, in which the fluctuation emf's are taken into account (Fig. 4). The generators of emf ε_{01} and ε_{02} represent the thermal noise of the walls of the first and second resonators, and ε_{B1} and ε_{B2} represent the noise of the active material in them. Kirchhoff's equations for the circuit of Fig. 4 at the resonant frequency can be written in the form

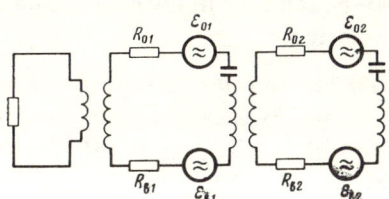

Fig. 4. Fluctuation emf's in the equivalent circuit of a two-resonator maser.

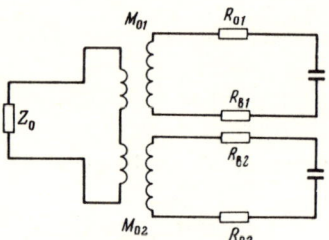

Fig. 5. Equivalent circuit of a maser with parallel connected resonators.

$$0 = Z_0 I_0 + j\omega M_{01} I_1,$$
$$\varepsilon_{01} + \varepsilon_{B1} = I_1(R_{01} - R_{B1}) + j\omega M_{01} I_0 + j\omega M_{12} I_2, \qquad (4.69)$$
$$\varepsilon_{02} + \varepsilon_{B2} = I_2(R_{02} - R_{B2}) + j\omega M_{12} I_1,$$

where I_0, I_1, and I_2 are the currents in the load, the first resonator, and the second resonator, respectively.

Hence, in the important practical case when $Q_{B1}^0 = Q_{B2}^0 = Q_B^0$, $\varkappa = 1/Q_B^0$, $T_{01} = T_{02} = T_0$, $T_{eff\,1} = T_{eff\,2} = T_{eff}$ and $Q_0 = Q_{01} = Q_{02} \gg Q_B^0$ the spectral density of the output noise equals

$$P_{\nu\,out} = 2\left(\frac{2/Q_{CB}}{2/Q_{CB} - 2/Q_B^0}\right)^2 \left(\frac{Q_{CB}}{Q_0}\frac{h\nu}{e^{h\nu/kT_0}-1} + \frac{Q_{CB}}{Q_B^0}\frac{h\nu e^{h\nu/k\,|T_{eff}|}}{e^{h\nu/k\,|T_{eff}|}-1}\right). \qquad (4.70)$$

Since for a two-resonator maser $Q_{CB} = \frac{1}{2}\frac{G_0^{1/2}-1}{G_0^{1/2}+1}Q_B^0$, then referred to the input, the spectral density of the inherent noise of this maser is given by formula (4.41), i.e., as might have been expected, it is the same as the spectral density of the input noise of a single-resonator maser.

Besides two-resonator masers with series coupled resonators, it is also possible to have a maser with parallel connected resonators. Resonators connected in this way in the feeder line can be particularly convenient in the centimeter band, where they were first used [50]. From the equivalent circuit of Fig. 5 for the case of a common primary winding of the coupling transformer, we can obtain the following expression for the parameter β:

$$\beta = \left\{\frac{1}{Q_{CB1}}\left[\frac{1}{Q_{02}} - \frac{1}{Q_{m2}} + i\left(X_2 + \frac{Q_l}{Q_{m2}}X_l\right)\right] + \frac{1}{Q_{CB2}}\left[\frac{1}{Q_{01}} - \frac{1}{Q_{m1}} + i\left(X_1 + \frac{Q_l}{Q_{m1}}X_0\right)\right] - \frac{2j\varkappa}{\sqrt{Q_{CB1}Q_{CB2}}}\right\}\left\{\left[\frac{1}{Q_{01}} - \frac{1}{Q_{m1}} + i\left(X_1 + \frac{Q_l}{Q_{m1}}X_l\right)\right]\left[\frac{1}{Q_{02}} - \frac{1}{Q_{m2}} + i\left(X_2 + \frac{Q_l}{Q_{m2}}X_l\right)\right] - \varkappa^2\right\}^{-1}, \qquad (4.71)$$

where Q_{CB1} and Q_{CB2} are the coupling Q-factors of the first and second resonators with the line, and $\varkappa$ is the coupling coefficient between the resonators. A knowledge of the parameter β enables us to obtain an expression for the frequency characteristic of the maser in any concrete case. It is interesting that when there is no cross coupling between the resonators, i.e., when $\varkappa = 0$, the two-frequency characteristic obtained in this way coincides with the frequency characteristic of a two-resonator maser with series resonators provided that the parallel resonators are symmetrically detuned by an amount 2Ω, i.e., if $X_1 = X_l + \Omega$, and $X_2 = X_l - \Omega$. In this case the relative detuning plays the part of the coupling coefficient from formulas (4.64)–(4.68). Naturally, the conclusions concerning the value of the instability and the estimate of the spectral density of the input noise which hold for a two-resonator maser with series connected resonators are also true for the case of parallel connection.

We will now turn to formulas (4.61) and (4.62) for the parameter β of two-resonator masers with series connected resonators. Formula (4.61) was obtained for the case of the same tuning of both resonators. However, under practical experimental conditions the frequency characteristic of two-resonator masers is formed not only by the choice of the optimum

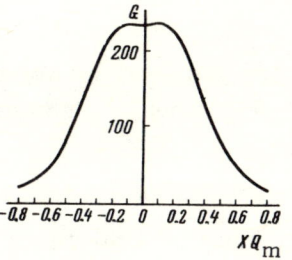

Fig. 6. Frequency characteristic of a two-resonator maser.

coupling between the resonators, but also by the detuning of the resonators with respect to one another. It is obvious that detuning of the resonators leads to a widening of the bandwidth. If the relative detuning of the resonators equals 2Ω and if it is symmetrical with respect to the line frequency $X_l = X$, formula (4.62) for the parameter β can be written in the form

$$\beta = \frac{1/Q_{CB}}{1/Q_{01} - 1/Q_{m1} + i(1 - \Omega + Q_l/Q_{m1})X + \dfrac{\varkappa^2}{1/Q_{02} - 1/Q_{m2} + i(1 + \Omega + Q_l/Q_{m2})X}}. \tag{4.72}$$

Hence we have that for two active resonators and with the assumption that $Q_{m1} = Q_{m2} \equiv Q_m \ll Q_{01} = Q_{02}$ the frequency characteristic is given by the formula

$$G(X) = \frac{\left[\dfrac{1}{Q_m Q_{CB}} + \varkappa^2 + \dfrac{1}{Q_m^2} + \Omega^2 - \left(1 + \dfrac{Q_l}{Q_m}\right)^2 X^2\right]^2 + \left[\left(1 + \dfrac{Q_l}{Q_m}\right)X\left(\dfrac{1}{Q_{CB}} + \dfrac{2}{Q_m}\right) - \dfrac{\Omega}{Q_{CB}}\right]^2}{\left[\dfrac{1}{Q_m Q_{CB}} - \varkappa^2 - \dfrac{1}{Q_m^2} - \Omega^2 + \left(1 + \dfrac{Q_l}{Q_m}\right)^2 X^2\right]^2 + \left[\left(1 + \dfrac{Q_l}{Q_m}\right)X\left(\dfrac{1}{Q_{CB}} - \dfrac{2}{Q_m}\right) - \dfrac{\Omega}{Q_{CB}}\right]^2}. \tag{4.73}$$

We can obtain a fairly rectangular frequency characteristic by suitable choice of the values of $\varkappa$ and Ω. For illustration, Fig. 6 shows the frequency characteristic of a two-resonator maser obtained using formula (4.73) for the values $Q_l = 0.3 Q_B^0$, $1/Q_{CB} = 2.02/Q_B^0$, $\varkappa^2 = 1/Q_B^{02}$ and $\Omega = 0.13/Q_B^0$. Note that taking into account the dependence of Q_B on the detuning relative to the central frequency of the line enables us to increase the slope of the sides of the frequency characteristic, which leads in a number of cases to larger "rectangularity."

Here it is convenient to consider one form of instability of the parameters of masers. When masers are used in radio receivers, in addition to phase and amplitude instabilities there can also be a change in the shape of the frequency characteristics of the masers. Such a change is possible when the matching of the antenna-feeder channels of the device changes. In radio astronomy, particularly when the spectral lines of cosmic radiation are being investigated, this leads to parasitic effects.

For ideal traveling-wave masers a mismatch of the channels has no effect on the frequency characteristic of the maser. For reflection resonator masers with circulators the effect of a mismatch can be taken into account by representing the antenna-feeder channel in the form of a long uniform line with an equivalent impedance Z_l. Assuming the existence of reflections from the input of the line

$$Z_l = Z_0 \frac{1 + \Gamma_l}{1 - \Gamma_l}, \tag{4.74}$$

where Z_0 is the wave impedance of the uniform line and Γ_l is the reflection coefficient from the input of the line. The presence of a circulator with a power decoupling α^2, which does not introduce additional phase shifts and which is matched on the right and on the left, is taken into account by setting $|\Gamma_l|^2 = \alpha^2 |\Gamma_a|^2$, where Γ_a is the reflection coefficient from the antenna. The quantity Γ_a is complex. We will assume that $\Gamma_l = \alpha \Gamma_a$. Then for a good circulator the modulus of Γ_l is small and

$$Z_l \approx Z_0(1 + 2\alpha\Gamma_a). \tag{4.75}$$

The gain of reflection masers is determined by the coupling Q-factor, which enters into the formula for the parameter $\beta = Z/Z_l$. In view of (4.75), when determining the frequency characteristics of masers we must assume that

$$1/Q_{\text{CB}} = 1/Q_{\text{CB}}^0 (1 - a - jb), \tag{4.76}$$

where Q_{CB} is the coupling Q-factor with an ideally matched line with a wave impedance Z_0, and the small corrections a and b are proportional to the real and imaginary parts of the reflection coefficient Γ_a: $a = 2\alpha \operatorname{Re} \Gamma_a$ and $b = 2\alpha \operatorname{Im} \Gamma_a$.

Consideration of the complex form of $1/Q_{\text{CB}}$ in formula (4.76) leads to the following conclusions for a single-resonator maser. When $b = 0$ the effect of a is to vary the amplification without detuning. For positive values of a the amplification increases, while for negative values of a the amplification falls. When $b \neq 0$, as well as a change in amplification, some retuning of the maser from the left or the right, depending on the sign of b, is also observed.

For a two-resonator maser with two active resonators the effect of mismatch leads mainly to a change in amplification. Sharp deformations of the frequency characteristic are not observed. However, when there is a detuning Ω, the effect of a mismatch leads to the appearance of separate peaks on the frequency characteristic and the curve ceases to be smooth. When $X = X_l$, $X_1 = X_l + \Omega$, $X_2 = X_l - \Omega$ and assuming $Q_{m1} = Q_{m2} \equiv Q_m \ll Q_{01} = Q_{02}$ the frequency characteristic for this case is given by the formula

$$G(X) = \Bigg\{ \Bigg[\frac{1-a}{Q_m Q_{\text{CB}}^0} + \varkappa^2 + \Omega^2 + \frac{1}{Q_m^2} + \frac{1}{Q_{\text{CB}}^0}\Big(\Omega - \Big(1 + \frac{Q_l}{Q_m}\Big)X\Big)b - $$

$$- \Big(1 + \frac{Q_l}{Q_m}\Big)^2 X^2 \Bigg]^2 + \Bigg[\Big(\frac{1-a}{Q_{\text{CB}}^0} + \frac{2}{Q_m}\Big)\Big(1 + \frac{Q_l}{Q_m}\Big)X - \frac{1}{Q_{\text{CB}}^0}\Big(1 - a\Big)\Omega + $$

$$+ b\frac{1}{Q_m Q_{\text{CB}}^0} \Bigg]^2 \Bigg\} \Bigg\{ \Bigg[\frac{1-a}{Q_m Q_{\text{CB}}^0} - \varkappa^2 - \Omega^2 - \frac{1}{Q_m^2} + \frac{1}{Q_{\text{CB}}^0}\Big(\Omega - \Big(1 + \frac{Q_l}{Q_m}\Big)X\Big)b + $$

$$+ \Big(1 + \frac{Q_l}{Q_m}\Big)^2 X^2 \Bigg]^2 + \Bigg[\Big(\frac{1-a}{Q_{\text{CB}}^0} - \frac{1}{Q_m}\Big)\Big(1 + \frac{Q_l}{Q_m}\Big)X - \frac{1}{Q_{\text{CB}}^0}(1-a)\Omega + b\frac{1}{Q_m Q_{\text{CB}}^0}\Bigg]^2 \Bigg\}^{-1}. \tag{4.77}$$

Figure 7 shows for illustration the frequency characteristic of a two-resonator maser constructed from formula (4.77) for the same values of the parameters which determine the amplification shown in the curve of Fig. 6, but taking into account the mismatch of the antenna channel: $a = 0.05$ and $b = 0$. The results of the calculations agree quite well with the experimental results, which show that the frequency characteristic of a two-resonator maser is in fact distorted when the matching of the antenna-feeder channel lessens. This distortion is larger the larger the amplification and the smaller the decoupling. The effect of a change in the VSWR of the channel from 1.1 to 1.7 for a gain of 20 dB is practically completely eliminated for a decoupling of 35 dB. One of the possible reasons for the change in the VSWR of the antenna channel is the change with time of the level of the liquid helium which fills the maser feeders. The dielectric constant of liquid helium is 1.048 and its effect cannot always be neglected.

§4. Multiresonator Masers

A further broadening of the bandwidth of resonator masers for a given gain can be obtained by using multiresonator masers. We will first consider the possibility of broadening the bandwidth of masers by the series connection of transmission resonator masers, completely decoupled from one another, for example, by using ferrite isolators. Such a construction is similar to an intermediate frequency multistage tube amplifier with the same tuned circuits in

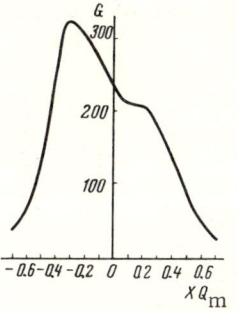

Fig. 7. Distorted frequency characteristic of a two-resonator maser.

the grid circuit of each stage. In this case the decoupling ensures that there is no feedback between the stages and that there is no possibility of the amplifier as a whole being unstable.

Figure 8 shows the equivalent circuit of one stage of the maser. By formulating Kirchhoff's equations for this circuit, we can obtain that the voltage transmission coefficient, assuming that the input and output coupling is the same and also that there are no inherent losses, is equal to

$$K = \frac{\omega^2 M^2/Z_0}{\omega^2 M^2/Z_0 + R_B + j\omega L \left(1 - \frac{\omega_0^2}{\omega^2}\right)}. \qquad (4.78)$$

Introducing in the usual way the coupling Q-factor and the Q-factor of the active material we obtain that when the resonator is tuned to the central frequency of the signal transition line

$$K = \frac{1/Q_{CB}}{1/Q_{CB} - 1/Q_m + jX\left(1 + \frac{Q_l}{Q_m}\right)}. \qquad (4.79)$$

Consequently, the frequency characteristic of one stage is given by the formula

$$G_1(X) = \frac{1/Q_{CB}^2}{(1/Q_{CB} - 1/Q_m)^2 + X^2\left(1 + \frac{Q_l}{Q_m}\right)^2}. \qquad (4.80)$$

Comparison of (4.80) with the expression (4.26) shows that the value of $G^{1/2}(0)\Delta\nu$ for a single-resonator transmission maser is in fact only half that of a reflection maser.

For the further analysis, using the connection between Q_B^0 and Q_{CB} shown in the gain at the resonance frequency

$$Q_{CB} = Q_B^0 \frac{G_1^{1/2}(0) - 1}{G_1^{1/2}(0)}, \qquad (4.81)$$

we can represent $G_1(X)$ in the form

$$G_1(X) = \frac{G_1(0)}{\left[1 + (\sqrt{G_1(0)} - 1)\frac{X^2 Q_l^2}{1 + X^2 Q_l^2}\right]^2 + X^2 Q_B^{02}(\sqrt{G_1(0)} - 1)^2 \left[1 + \frac{Q_l}{Q_B^0}\frac{1}{1 + X_l^2 Q_l^2}\right]^2}, \qquad (4.82)$$

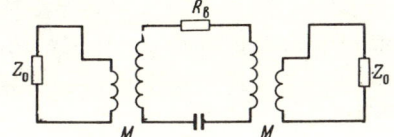

Fig. 8. Equivalent circuit of one stage of a multistage amplifier.

which is appropriate for a fixed value of Q_B^0. In the case of n completely decoupled stages the overall gain is given by

$$G(X) = G_1^n(X) = \frac{G_0}{\left\{\left[1 + \frac{\ln G_0}{2n} \frac{X^2 Q_l^2}{1+X^2 Q_l^2}\right]^2 + Q_B^{02} X^2 \left(\frac{\ln G_0}{2n}\right)^2 \left[1 + \frac{Q_l}{Q_B^0} \frac{1}{1+X^2 Q_l^2}\right]^2\right\}^n}, \quad (4.83)$$

where $G_0 = G_1^n(X)$ and we have assumed that for $n \gg 1$

$$G_0^{1/2n} - 1 \approx \frac{\ln G_0}{2n}.$$

In this case the bandwidth of the maser between the 3 dB levels is given by the equation

$$\left(1 + \frac{\ln G_0}{2n} \frac{X^2 Q_l^2}{1+X^2 Q_l^2}\right)^2 + X^2 Q_l^2 \left(\frac{\ln G_0}{2n}\right)^2 \left(\frac{1}{1+X^2 Q_l^2} + \frac{Q_B^0}{Q_l}\right)^2 = 2^{1/n}. \quad (4.84)$$

Confining ourselves to the frequencies which do not lie outside the limits of the line width, i.e., assuming that $X^2 Q_l^2 \ll 1$, we obtain a biquadratic equation for the detuning, the solution of which gives the connection between the bandwidth and the gain,

$$\Delta\nu \approx \Delta\nu_l \sqrt{\ln 2} / \sqrt{\ln G_0}, \quad (4.85)$$

which agrees with the similar formula for the traveling-wave maser considered earlier.

Therefore, as the number of stages increases for a constant value of the overall amplification, the bandwidth increases monotonically and in the limit tends to the bandwidth of a traveling-wave maser [51]. In this case a reduction in the gain of a single stage for a fixed value of the Q-factor of the active material is achieved by a reduction in the coupling Q-factor of this stage with the line.

However, in the case when $Q_l \ll Q_{CB}$, Q_B^0 and the bandwidth of the maser is determined by the bandwidth of the loaded resonator $\Delta\nu_H$, there is an optimum number of stages. In this case formula (4.80) is conveniently represented in the form

$$G_1(X) = \frac{G_{01}}{1 + X^2 Q_{CB}^2 G_{01}}. \quad (4.80a)$$

Then

$$G(X) = \frac{G_0}{(1 + X^2 Q_{CB}^2 G_0^{1/n})^n} \quad (4.81a)$$

and the bandwidth is determined by the equation

$$G_0^{1/n} X^2 Q_{CB}^2 = 2^{1/n} - 1 \approx \frac{\ln 2}{n}. \quad (4.82a)$$

Applying to (4.82a) the well-known method [52] of discussing intermediate frequency single-tuned circuit multistage amplifiers, we obtain that the bandwidth of the maser is a maximum when $n = \ln G_0$. Then the connection between the bandwidth and the gain is given by the relation

$$\Delta\nu = \Delta\nu_H \frac{\sqrt{\frac{\ln 2}{e}}}{\sqrt{\ln G_0}} \approx \frac{\Delta\nu_H}{\sqrt{n}}. \quad (4.83a)$$

An analysis [51] of multiresonator transmission masers with partially coupled resonators leads to the conclusion that there is an optimum value of the mutual decoupling, for which the maximum broad and uniform frequency characteristic is attained.

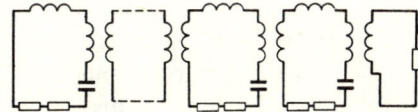

Fig. 9. Equivalent circuit of a multiresonator reflection maser.

It is obvious that a further broadening of the bandwidth of multistage masers can be achieved by detuning the various stages with respect to one another.

It is clear that the best use of the active material can be achieved in reflection multiresonator masers. Using the equivalent circuit of Fig. 9 it is easy to determine the impedance of a chain of n coupled circuits at the terminals of the input coupling transformer. This impedance is given by the continued fraction

$$Z = \cfrac{\omega^2 M_{01}^2}{R_{01} + R_{B1} + j\omega L_1 \left(1 - \cfrac{\omega_{01}^2}{\omega^2}\right) + \cfrac{\omega^2 M_{12}^2}{R_{02} + R_{B2} + j\omega L_2 \left(1 - \cfrac{\omega_{02}^2}{\omega^2}\right) + \cfrac{\omega^2 M_{23}^2}{R_{03} + R_{B3} \cdots}}} . \quad (4.86)$$

After introducing the corresponding Q-factors and the coupling coefficients we obtain the following expression for the parameter β, which determines the gain of the reflection maser:

$$\beta = \cfrac{1/Q_{CB}}{\cfrac{1}{Q_{01}} - \cfrac{1}{Q_{m1}} + i\left(X_1 + \cfrac{Q_l}{Q_{m1}} X_l\right) + \cfrac{\varkappa_{12}^2}{\cfrac{1}{Q_{02}} - \cfrac{1}{Q_{m2}} + i\left(X_2 + \cfrac{Q_l}{Q_{m2}} X_l\right) + \cfrac{\varkappa_{23}^2}{\cfrac{1}{Q_{03}} + \cfrac{1}{Q_{m3}} \cdots}}} . \quad (4.87)$$

The value of β enables us, using the relation $G = |(\beta - 1)|(\beta + 1)|^2$, to determine the gain and bandwidth in any concrete case.

Therefore, the use of multiresonator systems enables is to increase the bandwidth of masers considerably. Here, however, we must emphasize that when constructing multiresonator masers it is necessary to carry out an analysis of the stability of the system. As distinct from the case of completely decoupled multiresonator transmission masers, the reflection maser is looped by many feedbacks, which may cause self-excitation at frequencies lying outside the band of amplified frequencies, but within the limits of the signal transition resonance line. If we pass from the continued fraction of (4.87) to the characteristic equations, which involves quite lengthy reduction, we can use the Routh criterion to analyze the stability. It is obvious that the frequency dependence of the real part of the Q-factor of the active material (4.3) increases the stability margin of the device.

To conclude this section we will compare the various types of masers described above.

In principle, traveling-wave masers have the best characteristics. Their large bandwidth, stability, and, as will be seen later, their better amplitude and relaxation characteristics, and also the possibility of electronic retuning, at least over a certain limited range, are undoubtedly extremely attractive properties.

The nonreciprocal decoupling elements of the traveling-wave maser are always cooled to the temperature of the working material. This leads to an increase in the sensitivity in comparison with the case when one uses masers with circulators, which are at room temperature. The drawbacks of traveling-wave masers are the constructional complexity of the traveling-wave systems, the need to produce a fairly uniform magnetic field over large volumes, the large linear dimensions of the working material, and the considerable power of the pumping radiation. The latter leads to higher requirements regarding the cooling of the working material and for a given pumping power reduces the margin as regards the value of the saturation factor of the pumping transition. One must, however, bear in mind the extremely high operating reliability and simplicity of traveling-wave masers. Once they are constructed they do not require any adjustment and they operate very stably. Traveling-wave masers have obtained the widest use in the 3-15 cm wavelength band in the broadband receiving systems of long-distance communication links, in radio telescopes, and in radar.

In those cases when wide bandwidths are not required, the simpler single-resonator masers are used. Multiresonator systems, however, have undoubtedly better prospects. As far as broadband properties are concerned they can be close to traveling-wave masers, they are extremely stable, and are convenient for constructing receivers with fixed tuning, particularly in those wavebands in which construction of traveling-wave systems leads to very large dimensions. For wavelengths exceeding 20 cm, slow-wave structures for traveling-wave masers are difficult to realize in practice.

Several of the advantages of resonator masers are due to the fact that the volume of the working material of these masers is considerably less than in the case of traveling-wave masers. Therefore, a uniform magnetic field is required over a smaller volume, which reduces its weight and size. The smaller power of the pumping radiation leads to less evaporation of the cooling agent, a larger duration of continuous operation, and easier requirements as regards the closed-cycle cooling system. In addition, one must bear in mind that technologically it is much easier to manufacture uniform crystals of small dimensions.

We note once again that two-resonator masers with small dimensions are sufficiently broadband and extremely stable. Amplifiers of this type have performed well in operation [53, 54].

All these advantages of resonator masers can be particularly important in the decimeter band. On the other hand, in the short-wave part of the centimeter and in the long-wave part of the millimeter bands, in view of the increase in the technical difficulty of constructing slow-wave systems for traveling-wave masers, it may be more advantageous to use resonator masers, the bandwidth of which for reasonable gains does not differ very much from the bandwidth of traveling-wave masers.

At very short wavelengths, in the millimeter and submillimeter bands, difficulties arise in constructing single-mode high-Q resonators filled with the working material. Therefore, it is better in these wavebands to construct traveling-wave masers, matched to the feeders, or regenerative masers with partially reflecting mirrors.

CHAPTER V

Nonlinear Effects in Masers

Masers are linear and are characterized by a complex frequency transmission function. The discussion carried out in the previous sections was devoted to determining and analyzing the frequency transmission functions of masers of different types. The initial parameters of these transmission functions to be determined are the properties of the complex Q-factor of the active material. However, masers are linear only for small input signal levels, when the Q-factor of the active material does not depend on the signal intensity. However, due to the effect of saturation when the signal intensity is increased the population difference of the signal transition falls, the gain is reduced, and the linearity of the maser is destroyed. The nonlinear properties of masers require special consideration. Strictly speaking, when the linearity breaks down it is not possible to use the Q-factor of the active material to determine the frequency transmission function. However, in the case of saturation by a monochromatic signal at the resonance frequency, it is easy to obtain the dependence of the population difference on the signal intensity and we can formally assume that this population difference determines the value of the Q-factor at the resonance frequency for a given signal level. The value of the Q-factor obtained in this way can be used to determine the dependence of the gain on the input signal power and to determine the relaxation time of the maser [38, 55].

It is clear that such a description can be used to analyze the transmission through masers of slowly varying strong signals. At the same time the width of the spectrum of the signals is often close to the bandwidth of the maser. To analyze such cases it is necessary to solve the straightforward problem of the interaction of the strong radiation field with the active material of a three-level paramagnetic maser [56].

This chapter is devoted to a consideration of these problems.

§1. Saturation and Restoration of the Q-Factor of the Active Material

We will consider the dependence of the Q-factor of the active material at the center of the signal transition line on the intensity of the amplified signals. In this and the later discussion we will neglect quantities of the order of T_2/T_1. We are interested in the Q-factor Q_B^0. In accordance with the formula (3.38) the quantity $1/Q_B^0$ may vary under the influence of the signal intensity only if the inversion coefficient I changes. The reduction in inversion for an increase in the signal can be determined with the help of the velocity equation (3.15), in which we additionally take into account the probability of transitions induced by the signal

$$-(w_{12} + w_{13})\rho_{11} + w_{12}\rho_{22} + w_{31}\rho_{33} - (\rho_{11} - \rho_{33})W_p = 0,$$
$$w_{12}\rho_{11} - (w_{21} + w_{23})\rho_{22} + w_{32}\rho_{33} - (\rho_{22} - \rho_{33})W_{sig} = 0, \quad (5.1)$$
$$\rho_{11} + \rho_{22} + \rho_{33} = 1,$$

where the probability of induced transitions W_p and W_{sig} are determined by the intensity of the pumping and signal fields

$$W_p = \frac{|\mu_{31}|^2}{\hbar^2} T_2 H_p^2 \quad \text{and} \quad W_{sig} = \frac{|\mu_{32}|^2}{\hbar^2} T_2 H_{sig}^2.$$

For large intensities of the pumping radiation when fairly deep saturation of the pumping transition is obtained, from (5.1) it is easy to obtain

$$\Delta_{32} = \frac{w_{21} - w_{12} - (w_{32} - w_{23})}{2w_{21} + w_{12} + w_{32} + 2w_{23} + 3W_{sig}}. \quad (5.2)$$

Comparing (5.2) with formulas (3.17), (3.19), and (3.38), we can write that

$$Q_B^0 = Q_B^0(W_{sig} = 0)\{1 + W_{sig}T_{1sig}\}, \quad (5.3)$$

where $Q_B^0(W_{sig} = 0)$ denotes the value of the Q-factor of the active material at the center of the line for vanishingly small signals, and T_{1sig} is a certain effective spin-lattice relaxation time for the signal transition

$$T_{1sig} = \frac{3}{2w_{21} + w_{12} + w_{32} + 2w_{23}}. \quad (5.4)$$

This time differs somewhat from the time T_1, introduced in formula (3.18). In the radio band, and assuming the equality of the partial probabilities w_{ik} to one another, the time T_{1sig} is $8/3$ times larger than the time T_1. Therefore, for signals of large intensity, in view of the effect of saturation of the signal transition the inversion is considerably reduced.

It is of considerable interest to determine the characteristic restoration time of the inversion after removing the field of the saturating signal. It is to be expected that this time will essentially be the same as the spin-lattice relaxation time. Equations (5.1) which describe the stationary conditions must be supplemented by the corresponding derivatives. As a result we can obtain that for conditions of strong saturation of the pumping transition the inversion is restored to its nominal value exponentially

$$1/Q_B^0 = \frac{1}{Q_B^0(W_{sig}=0)} (1 - e^{-t/T_{1\,rest}}) \qquad (5.5)$$

with a characteristic time

$$T_{1\,rest} = \frac{2}{w_{12} + w_{21} + w_{13} + w_{31} + w_{23} + w_{32}}. \qquad (5.6)$$

In the particular case when all the relaxation probabilities are equal to one another the time $T_{1\,sig}$ is 1.5 times larger than the time $T_{1\,rest}$. For estimates we can take the times T_1, $T_{1\,sig}$, and $T_{1\,rest}$ as being practically equal. In what follows we will use the symbol T_1 for all these times.

Therefore, the spin-lattice relaxation time determines both the level of the signal power which saturates the Q-factor of the active material and the rate at which it is restored after this signal is removed. In this case we must not forget that the time T_1 also determines the value of the power of the pumping radiation necessary to obtain inversion [see formula (3.50)].

§2. Saturation of Masers

We will consider a traveling-wave maser. The flow of energy P of the traveling wave which propagates along the slow-wave structure is determined by the value of the attenuation factor (or the gain)

$$\frac{dP}{dX} = 2\alpha P, \qquad (5.7)$$

where α is given by formula (4.1). Neglecting the ohmic losses and considering only the central frequency, we can write

$$\frac{dP}{dX} = \frac{2\pi\nu}{uQ_B^0} P, \qquad (5.8)$$

where Q_B^0 is given by formula (5.3), i.e., it depends on P. The calculation of the dependence of Q_B^0 on P requires fixing the type of oscillations and the distribution of the field in the traveling-wave slow-wave structure. We will make an approximate estimate connecting the linear energy density P/u with the mean value, over the perpendicular cross section of the system, of the intensity of the magnetic field of the signal H_{sig}^2:

$$H_{sig}^2 = 4\pi P/\Omega u, \qquad (5.9)$$

where Ω is the area of cross section. Formulas (5.3) and (5.9) enable us to write equation (5.7) in the form

$$\frac{dP}{dX} = \frac{\alpha_0 P}{1 + SP}, \qquad (5.10)$$

where $\alpha_0 = \dfrac{2\pi\nu}{uQ_B^0(W_{sig}=0)}$ is the gain for small input signals, and S is the saturation factor of the signal transition for a traveling-wave maser, and equals

$$S = \frac{|\mu_{32}|^2}{\hbar^2} T_2 T_1 \frac{4\pi}{\Omega u}. \tag{5.11}$$

Equation (5.10) is easily integrated:

$$P(X) = P_{in} e^{\alpha_0 X} e^{-S[P(X)-P_{in}]}. \tag{5.12}$$

Consequently, the gain G of a traveling-wave maser, equal by definition to the ratio $P(l)/P_{in}$, where l is the length of the maser, is connected with the power of the input signal by the equation

$$G = G_0 e^{-(G-1)SP_{in}}. \tag{5.13}$$

Here G_0 is the gain for vanishingly small input signals. Hence, we see that as the input power is increased G tends to one. Expression (5.11) for the saturation factor can be written in a form which is more convenient for discussion. Substituting in (5.11)

$$u = 2\pi\nu/Q_B^0 (w_{sig} = 0) \ln G_0, \qquad |\mu_{in}|^2 T_2/\hbar^2 = W_{sig}/\bar{H}_{sig}^2$$

and

$$Q_B^0 (W_{sig} = 0) = \nu \bar{H}_{sig}^2 V_{per}/4P_{rad},$$

where $P_{rad} = \eta V_{per} W_{sig} h\nu \Delta n$, V_{per} is the volume of one spatial period of the slow-wave structure, η is the filling factor, and Δn is the particle density in the upper level, we obtain

$$S = \frac{T_1 \ln G_0}{2\eta V \Delta n h\nu} = \frac{T_1}{2\varepsilon} \ln G_0. \tag{5.14}$$

Here, $V = \Omega l$ is the volume of the traveling-wave system. We see from this notation that T_1/S essentially has the meaning of the energy ε, stored in the system of inverted particles.

Therefore, the saturation factor of the signal transition decreases with an increase in the number of active particles and the energy of a quantum. The saturation factor increases with increase in gain, but for a traveling-wave maser not too sharply.

The relation between the gain of traveling-wave masers and the input signal power is shown in Fig. 10 for amplifiers with $G_0 = 1000$, $G_0 = 100$, and $G_0 = 10$. We see from Fig. 10 that the effect of an increase in the signal power occurs very slowly at first. On each of these curves there is a part for which a ten-fold increase in the signal power changes the amplification by less than 1 dB. We also see that amplifiers with larger G_0 are more saturated by signals that are strong relative to G_0, than are amplifiers with smaller G_0. When $SP_{in} > 1$ the gain practically disappears. This indicates that it disappears when the energy carried by the signal in a time T_1 becomes comparable with the energy stored in the system of inverted particles.

We will now consider a reflection single-resonator maser. If we assume that the Q-factor of the resonator Q_0 considerably exceeds the coupling Q-factor Q_{CB}, then in accordance with (4.27) at the resonance frequency

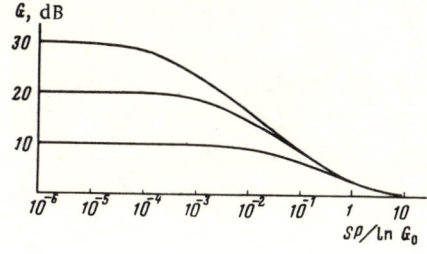

Fig. 10. Dependence of the gain of traveling-wave masers on the input signal power.

$$G^{1/2} = \frac{Q_B^0 + Q_{CB}}{Q_B^0 - Q_{CB}}. \tag{5.15}$$

Taking (5.3) into account formula (5.15) takes the form

$$G^{1/2} = \frac{G_0^{1/2} + \frac{1}{2}(G_0^{1/2} + 1) W_{sig} T_1}{1 + \frac{1}{2}(G_0^{1/2} + 1) W_{sig} T_1}, \tag{5.16}$$

where G_0 is the gain for small input signals. For a large signal intensity, i.e., for $W_{sig} T_1 \gg 1$, $G^{1/2} \to 1$.

As in the previous case, to calculate the dependence of the gain on the signal power it is necessary to know the distribution of the high-frequency fields in the resonator. However, for estimates it is sufficient to relate the power P_{rad} radiated by the active material in terms of the Q-factor Q_B^0 with the magnetic energy density $\overline{H_{sig}^2}$ averaged over the volume of the resonator V

$$\overline{H_{sig}^2} = \frac{4 Q_B^0}{\nu} \frac{P_{rad}}{V}. \tag{5.17}$$

Since

$$Q_B^0 = \frac{G^{1/2} + 1}{G^{1/2} - 1} Q_{CB} = \frac{G_0^{1/2} - 1}{G_0^{1/2} + 1} \frac{G^{1/2} + 1}{G^{1/2} - 1} Q_B^0 (W_{sig} = 0)$$

and from the condition for energy balance $P_{rad} + P_{in} = P_{out}$ it follows that $P_{rad} = (G-1)P_{in}$, and (5.17) can be written in the form

$$\overline{H_{sig}^2} = 4 Q_B^0 (W_{sig} = 0) \frac{G^{1/2} + 1}{\nu V} P_{in} \frac{G_0^{1/2} - 1}{G_0^{1/2} + 1}. \tag{5.18}$$

Substituting in (5.16) $W_{sig} = |\mu_{32}|^2 \overline{H_{sig}^2} T_2 / \hbar^2$, we obtain, taking into account (5.18), an equation which connects the gain of a single-resonator maser with the input signal power

$$G_2^{1/2} = \frac{G_0^{1/2} + (G^{1/2} + 1)^2 S P_{in}}{1 + (G^{1/2} + 1)^2 S P_{in}}, \tag{5.19}$$

where S is the saturation factor of the signal transition, given by

$$S = 2 \frac{|\mu_{32}|^2}{\hbar^2} T_2 T_1 \frac{Q_B^0 (W_{sig} = 0)}{\nu V} (G_0^{1/2} - 1). \tag{5.20}$$

The relationship obtained is shown in Fig. 11. As in the case of a traveling-wave maser it is convenient to give a somewhat different form to the expression for the saturation factor. Carrying out a conversion similar to the one employed in the case of a traveling-wave maser, we obtain

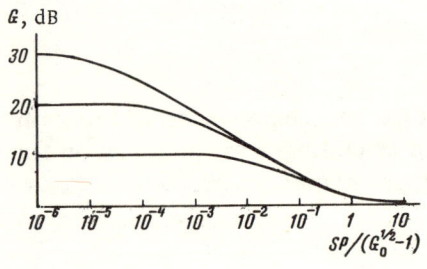

Fig. 11. Gain of a single-resonator maser as a function of the input signal power.

$$S = \frac{T_1 (G_0^{1/2} - 1)}{2\eta V \Delta n h \nu} = \frac{T_1}{2E} (G_0^{1/2} - 1). \tag{5.21}$$

In view of the regenerative nature of a resonator maser its saturation factor depends more sharply than in the case of a traveling-wave maser on the initial gain G_0. Other things being equal, the dynamic range of a traveling-wave maser for a gain of 20 dB is twice as large, and for a gain of 30 dB is five times as large as the dynamic range of a resonator maser. From Figs. 10 and 11 we also see that although the curves of the dependence of the gain of a resonator maser and a traveling-wave maser on SP_{in} do not differ very much from one another, under traveling-wave conditions the initial drop in gain occurs less sharply. To estimate the order of magnitude of the saturation factors we will make a numerical estimate of the product $2\eta V \Delta n h\nu$. In the three-centimeter band for $\eta = 1$, $V = 1$ cm^3, $I = 2$, $N = 2 \cdot 10^{19}$ cm^{-3}, the value of $2\eta V \Delta h\nu$ is approximately $1.5 \cdot 10^{-5}$ J. For $T_1 = 0.1$ sec and $G_0 = 10^2$ the saturation factor for a single-resonator maser is $0.6 \cdot 10^{-5}$ W^{-1}.

The time of restoration of the gain after removing the saturating signal is determined by the speed of the spin-lattice relaxation processes. In view of the regenerative nature of masers this time exceeds T_1. Taking into account (5.5), we obtain for a single-resonator maser

$$G^{1/2}(t) = \frac{2G^{1/2} - (G_0^{1/2} - 1) e^{-t/T_1}}{2 + (G_0^{1/2} - 1) e^{-t/T_1}}. \tag{5.22}$$

Hence, the restoration time of the gain τ_α from $G^{1/2} = 1$ to $G^{1/2} = \alpha G_0^{1/2}$ equals

$$\tau_\alpha = T_1 \ln \frac{(G_0^{1/2} - 1)(\alpha G_0^{1/2} + 1)}{2 G_0^{1/2} (1 - \alpha)}. \tag{5.23}$$

For $\alpha = 0.63$ [restoration to the level which is less than the maximum by a factor of $e/(e-1)$] and $G_0^{1/2} = 10$, $\tau_\alpha = 2.2 T_1$. For a traveling-wave maser the use of expression (5.5) leads to the relation

$$\frac{\ln G_0 - \ln G}{\ln G_0} = e^{-t/T_1}, \tag{5.24}$$

which gives a restoration time to the value $G = \alpha^2 G_0$

$$\tau_\alpha = T_1 \ln \frac{\ln G_0}{\ln 1/\alpha^2}. \tag{5.25}$$

For $\alpha = 0.63$ and $G_0 = 100$, $\tau_\alpha = 1.6 T_1$. The restoration time of a traveling-wave maser is less than for a single-resonator maser. The difference between them increases as G_0 is increased. Therefore, consideration of the amplitude and time characteristics of masers shows that the spin-lattice relaxation time T_1 essentially determines these characteristics. An increase in T_1 leads not only to an easing of the saturation conditions as regards the pumping transition which facilitates the attainment of inversion and increases the stability, but also leads to a reduction in the dynamic range and to an increase in the restoration time. In certain applications, for example, in radio astronomy or planetary radar, this fact is not important and we must give preference to masers whose working material possesses large T_1, since deeper saturation of the pumping transition leads to large stability of the quantity Q_B^0, and the larger value of the ratio T_1/T_2 strongly reduces the effect of unstable pumping.

In other applications, for example, in radar or in communication links, where rapid fadings are possible, it is better to use a material with a short time T_1. As an example of the effect of the finite restoration time of the maser we will consider the case of the possible application of a single-resonator maser to increase the range of a pulse radar. In this case the probing pulse saturates the maser, which must be able to recover before the pulse returns.

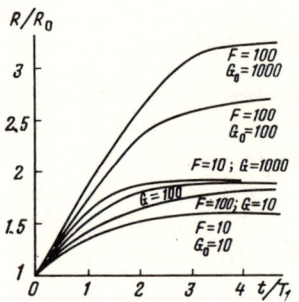

Fig. 12. Time dependence of the restoration of the range of action.

When the pulse is finished the process of gain restoration begins, described by formula (5.22). We can assume for the estimate that in accordance with the restoration of the gain the increase in sensitivity (2.15)-(2.18) is also restored. Combining these formulas with (5.22) and assuming that the range of the radar is proportional to the fourth root of the sensitivity, we can construct curves which show the time dependence of the restoration of the range of action of the radar after radiating the probing pulse which saturates the maser. Figure 12 shows curves constructed for a single-resonator maser for $T_b = 150°K$ and $\alpha = 0.5$. We recall that the time T_1 corresponds to a distance $R = cT_1/2$. Therefore, we see from Fig. 12 that when T_1 equals 1 msec a significant improvement in the range begins after a time interval which corresponds to a distance of 150 km.

§3. Polarization of the Active Material under the Action of the Sum of a Strong Monochromatic Signal and a Weak Signal

We have seen that the gain of a maser falls under the action of a sufficiently intense source. The dependence of the gain on the power of the input signal which appears in this case makes the maser nonlinear, which can hardly but have an effect when it is used for the reception of information. An analysis of the distortion of the amplitude-modulated signal, based on the use of the formulas of §2 of this chapter, was made in [57] for the case of slow modulation, which quasistationarily causes a change in gain. In cases when the modulation frequency is comparable with the rate of population relaxation or exceeds it, one cannot use the rate equations, and the formulas obtained in §2, Chapter V, are not applicable to the analysis of nonlinear distortions in masers.

As already pointed out at the beginning of this chapter, to analyze these cases when the width of the signal spectrum exceeds the rate of population relaxation, it is necessary to solve the direct problem of the interaction of a strong radiation field with the active material of the maser. In this case a convenient method is the determination of the polarization of the active material using the density matrix. This method was used in [58] for a theoretical investigation of such nonlinear effects as the generation of fields of the second harmonic and combination frequencies, which occur in traveling-wave masers. The discussion was limited to the case of two energy levels and was essentially carried out assuming the absence of saturation of the maser by the signal fields. In this approximation the intensities of the fields of the second harmonic and the combination frequencies turn out to be 9-10 orders less than the intensity of the initial fields at the input of the maser. Specially performed experiments [59-61] have verified this result. Distortion of the modulation was not considered in [58]. Much stronger nonlinear effects occur with saturation of the maser when the gain is reduced under the action of a fairly intense signal.

In the further discussion, when determining the nonlinear distortion in a traveling-wave maser when it is saturated by the field of a strong signal, we will use the results of the calculation of the polarization of a quantum system with three energy levels and with two relaxation times under the action of the sum of a weak field and a monochromatic strong field. The polarization is determined by means of the density matrix.

We will write once again the equations for the density matrix:

$$\dot{\rho}_{nm} = j\omega^0_{nm}\rho_{nm} + \frac{1}{\tau_{nm}}\rho_{nm} + j/\hbar \sum_k (\mu_{nk}\rho_{kn} - \rho_{nk}\mu_{kn}) H, \quad m \neq n,$$

$$\dot{\rho}_{nn} = -\sum_k (w_{nk}\rho_{nn} - w_{kn}\rho_{kk}) + j/\hbar \sum_k (\mu_{nk}\rho_{kn} - \rho_{nk}\mu_{kn}) H,$$

(5.26)

where, as distinct from the notation in § 2, Chapter III, the natural frequency of the transition between the levels m and n is denoted by ω^0_{nm}, and a magnetic field is assumed which contains frequencies close to the frequency ω^0_{31} (the field of the pumping radiation) and to the frequency ω^0_{32} (the signal field):

$$H = H_{31} \cos(\omega_{31}t + \varphi_{31}) + H_{32} \cos(\omega_{32}t + \varphi_{32}) + h \cos[(\omega_{32} + \Omega)t + \varphi].$$

(5.27)

Here H_{32} is the amplitude of the strong field, and h is the weak field tuned away from the strong field by a frequency Ω. The frequency $\omega_{32} + \Omega$ lies within the limits of the line width of the signal transition $3 \to 2$.

Introducing the new unknown functions

$$Z_{13} = -j \frac{|\mu_{13}|}{\mu_{13}} e^{-j\omega_{31}t - j\varphi_{31}} \rho_{13},$$

$$Z_{23} = -j \frac{|\mu_{23}|}{\mu_{23}} e^{-j\omega_{32}t - j\varphi_{32}} \rho_{23},$$

$$Z_{12} = \frac{\mu_{13}}{|\mu_{13}|} \frac{\mu_{23}}{|\mu_{23}|} e^{-j(\omega_{31}-\omega_{32})t - j(\varphi_{31}-\varphi_{32})} \rho_{12},$$

$$Z_1 = \rho_{33} - \rho_{11}, \quad Z_2 = \rho_{33} - \rho_{22}$$

(5.28)

and using the condition of conservation of the sum of the relative populations $\rho_{33} + \rho_{22} + \rho_{11} = 1$, we obtain from (5.26) a set of equations for the slow amplitudes of the nondiagonal elements Z_{ij} and of the population differences Z_1 and Z_2, which determine the polarization of the active material.

$$\dot{Z}_{12} + \gamma_{12} Z_{12} = \frac{1}{2} A_{31} Z^*_{23} + \frac{1}{2} A_{32} Z_{13} + \frac{1}{2} a e^{-j\Omega t - j\varphi + j\varphi_{32}} Z_{13},$$

$$\dot{Z}_{13} + \gamma_{13} Z_{13} = \frac{1}{2} A_{31} Z_1 - \frac{1}{2} A_{32} Z_{12} - \frac{1}{2} a e^{-j\Omega t + j\varphi - j\varphi_{32}} Z_{12},$$

$$\dot{Z}_{23} + \gamma_{23} Z_{23} = \frac{1}{2} A_{32} Z_2 - \frac{1}{2} A_{31} Z^*_{12} + \frac{1}{2} a e^{j\Omega t + j\varphi - j\varphi_{32}} Z_2,$$

$$\dot{Z}_1 + \Lambda_{11} Z_1 + \Lambda_{12} Z_2 = \Lambda_1 - A_{31}(Z_{13} + Z^*_{13}) - \frac{1}{2} A_{32}(Z_{23} + Z^*_{23}) - \frac{1}{2} a (Z_{23} e^{-j\Omega t - j\varphi + j\varphi_{32}} + Z^*_{23} e^{j\Omega t + j\varphi - j\varphi_{32}}),$$

$$\dot{Z}_2 + \Lambda_{21} Z_1 + \Lambda_{22} Z_2 = \Lambda_2 - A_{32}(Z_{23} + Z^*_{23}) - \frac{1}{2} A_{31}(Z_{13} + Z^*_{13}) - a (Z_{23} e^{-j\Omega t - j\varphi + j\varphi_{32}} + Z^*_{23} e^{-j\Omega t + j\varphi - j\varphi_{32}}).$$

(5.29)

Here we have kept the notation of § 2, Chapter III, for the quantities

$$\gamma_{nm} = \frac{1}{\tau_{nm}} + j(\omega_{nm} - \omega^0_{nm}) \text{ and } A_{nm} = \frac{|\mu_{nm}|}{\hbar} H_{nm},$$

and, in addition, we have also introduced the notation

$$\Lambda_{11} = \frac{2w_{12} + w_{21} + 2(2w_{13} + w_{31}) - (w_{23} - w_{32})}{3},$$

$$\Lambda_1 = \frac{w_{12} - w_{21} + 2(w_{13} - w_{31}) + (w_{23} - w_{32})}{3},$$

(5.30)

$$\Lambda_{12} = \frac{2(w_{31} - w_{13}) - (w_{12} + 2w_{21}) + 2w_{23} + w_{32}}{3},$$

$$\Lambda_{21} = \frac{-(2w_{12} + w_{21}) + (2w_{13} + w_{31}) - 2(w_{23} - w_{32})}{3},$$

$$\Lambda_{22} = \frac{w_{12} + 2w_{21} - (w_{13} - w_{31}) + 2(2w_{23} + w_{32})}{3}, \qquad (5.30)$$

$$\Lambda_{2} = \frac{w_{21} - w_{21} + w_{13} - w_{31} + 2(w_{23} - w_{32})}{3},$$

$$a = \frac{|\mu_{31}|}{\hbar} h.$$

On the right side of the equations obtained in this way we have retained only the quantities which oscillate with slow frequencies, i.e., with frequencies which are close to zero and which do not exceed the line width.

We shall look for the nonlinear effects which lead to the appearance, in particular, of combination frequencies. Therefore, we cannot assume that the amplitude of the oscillations of each nondiagonal element of the density matrix is constant, and the frequency coincides with the frequency of the induced field ω_{nm} corresponding to it. In view of this we will retain the derivatives $\dot{Z}_{nm}$ in equations (5.29).

By our assumption the field a is small and we will look for a solution of (5.29) in the form of the sum of an accurate solution X_n for $a = 0$ and a correction x_n due to the field a, i.e., in the form

$$Z_n = X_n + x_n.$$

The equations of the zero approximation have the form

$$\dot{X}_{12} + \gamma_{12}X_{12} - \tfrac{1}{2}A_{31}X_{23}^* - \tfrac{1}{2}A_{32}X_{13} = 0,$$
$$\dot{X}_{13} + \gamma_{13}X_{13} + \tfrac{1}{2}A_{32}X_{12} - \tfrac{1}{2}A_{31}X_1 = 0,$$
$$\dot{X}_{23}^* + \gamma_{23}^*X_{23}^* + \tfrac{1}{2}A_{31}X_{12} - \tfrac{1}{2}A_{32}X_2 = 0, \qquad (5.31)$$
$$\dot{X}_2 + \Lambda_{21}X_1 + \Lambda_{22}X_2 + \tfrac{1}{2}A_{31}(X_{13} + X_{13}^*) + A_{32}(X_{23} + X_{23}^*) = \Lambda_2,$$
$$\dot{X}_1 + \Lambda_{11}X_1 + \Lambda_{12}X_2 + A_{31}(X_{13} + X_{13}^*) + \tfrac{1}{2}A_{32}(X_{23} + X_{23}^*) = \Lambda_1.$$

Neglecting terms of the form ax we obtain for the corrections

$$\dot{x}_{12} + \gamma_{12}x_{12} - \tfrac{1}{2}A_{31}x_{23}^* - \tfrac{1}{2}A_{32}x_{13} = \tfrac{1}{2}e^{-j\Omega t - j\varphi + j\varphi_{32}}X_{13},$$
$$\dot{x}_{13} + \gamma_{13}x_{13} + \tfrac{1}{2}A_{32}x_{12} - \tfrac{1}{2}A_{31}x_1 = -\tfrac{1}{2}ae^{-j\Omega t - j\varphi + j\varphi_{32}}X_{12},$$
$$\dot{x}_{23}^* + \gamma_{23}^*x_{23}^* + \tfrac{1}{2}A_{31}x_{12} - \tfrac{1}{2}A_{32}x_2 = \tfrac{1}{2}ae^{j\Omega t + j\varphi - j\varphi_{32}}X_2, \qquad (5.32)$$
$$\dot{x}_1 + \Lambda_{11}x_1 + \Lambda_2x_2 + A_{31}(x_{13} + x_{13}^*) + \tfrac{1}{2}A_{32}(x_{23} + x_{23}^*) = \tfrac{1}{2}a(X_{23}e^{-j\Omega t - j\varphi + j\varphi_{32}} + X_{23}^*e^{j\Omega t + j\varphi - j\varphi_{32}}),$$
$$\dot{x}_2 + \Lambda_{21}x_1 + \Lambda_{22}x_2 + \tfrac{1}{2}A_{31}(x_{13} + x_{13}^*) + A_{32}(x_{32} + x_{23}^*) = -a(X_{23}e^{-j\Omega t - j\varphi + j\varphi_{32}} + X_{23}^*e^{j\Omega t + j\varphi - j\varphi_{32}}).$$

The equations of the zero approximation agree with the equations solved in Chapter III when calculating the complex Q-factor of the active material. Bearing in mind the stationary conditions, we will look for solutions of these equations (5.31) which do not depend on time. We will solve the set of equations (5.32) by making the substitutions

$$x_{nm} = x_{nm}^+ e^{j\Omega t + j\varphi - j\varphi_{32}} + x_{nm}^- e^{-j\Omega t - j\varphi + j\varphi_{32}},$$
$$x_n = x_n^+ e^{j\Omega t + j\varphi - j\varphi_{32}} + x_n^- e^{-j\Omega t - j\varphi + j\varphi_{32}}, \qquad (5.33)$$

i.e., we separate out in the correction terms x_{nm} and $\dot{x}_n$ the amplitudes $x_{nm}^{\pm}$ and $x_n^{\pm}$ which do not depend on time.

Without writing down in complete form the solutions of the set of linear algebraic equations obtained in this way, we will give the expressions for the quantities X_{23} and $x_{23}^{\pm}$ which we require.

The zero approximation is

$$X_{23} = \frac{1}{2} \frac{1}{\gamma_{23}} X_2,$$

$$X_2 = \left[\Lambda_{11}\Lambda_2 - \Lambda_{21}\Lambda_1 - \left(\Lambda_2 - \frac{\Lambda_1}{2}\right)\left(\frac{1}{\gamma_{13}^*} + \frac{1}{\gamma_{13}}\right)\frac{A_{31}^2}{2}\right] \times$$
$$\times \left[\Lambda_{11}\Lambda_{22} - \Lambda_{12}\Lambda_{21} + \left(\Lambda_{11} - \frac{\Lambda_{21}}{2}\right)\left(\frac{1}{\gamma_{23}} + \frac{1}{\gamma_{23}^*}\right)\frac{A_{32}^2}{2} + \left(\Lambda_{22} + \frac{\Lambda_{12}}{2}\right)\left(\frac{1}{\gamma_{13}^*} + \frac{1}{\gamma_{13}}\right)\frac{A_{31}^2}{2} + \right.$$
$$\left. + \frac{3}{16}\left(\frac{1}{\gamma_{13}^*} + \frac{1}{\gamma_{13}}\right)\left(\frac{1}{\gamma_{23}^*} + \frac{1}{\gamma_{23}}\right)A_{31}^2 A_{32}^2\right]^{-1}. \quad (5.34)$$

The correction is

$$x_{23}^+ = \frac{1}{\gamma_{23} + j\Omega}\left(\frac{1}{2} a X_2 + \frac{1}{2} A_{32} x_2^+\right),$$

$$x_{23}^- = \frac{1}{\gamma_{23} + j\Omega} \frac{1}{2} A_{32} x_2^-,$$

$$x_2^- = (x_2^+)^*, \quad (5.35)$$

$$x_2^+ = \frac{\left(\frac{1}{2}\frac{1}{\gamma_{23} + j\Omega} A_{32} a X_2 - a X_{23}^*\right)\left[\Lambda_{11} + j\Omega - \frac{\Lambda_{21}}{2} + \frac{3}{8} A_{31}^2\left(\frac{1}{\gamma_{13} + j\Omega} + \frac{1}{\gamma_{13}^* + j\Omega}\right)\right]}{F},$$

where the denominator F differs from the denominator of the expression for X_2 in the fact that the terms of the form Λ_{ik} and γ_{ik} which have the dimensions of frequency, have the additive correction $j\Omega$:

$$F = (\Lambda_{11} + j\Omega)(\Lambda_{22} + j\Omega) + \left(\Lambda_{12} + j\Omega - \frac{\Lambda_{21}}{2}\right)\left(\frac{1}{\gamma_{23} + j\Omega} + \frac{1}{\gamma_{23}^* + j\Omega}\right)\frac{A_{32}^2}{2} +$$
$$+ \left(\Lambda_{22} + j\Omega - \frac{\Lambda_{12}}{2}\right)\left(\frac{1}{\gamma_{13} + j\Omega} + \frac{1}{\gamma_{13}^* + j\Omega}\right)\frac{A_{31}^2}{2} + \frac{3}{16}\left(\frac{1}{\gamma_{13} + j\Omega} + \frac{1}{\gamma_{13}^* + j\Omega}\right)\left(\frac{1}{\gamma_{23} + j\Omega} + \frac{1}{\gamma_{23}^* + j\Omega}\right) A_{31}^2 A_{32}^2.$$

Formulas (5.34) and (5.35) are accurate to within quantities of the order of $A_{31}^2 \tau_{32}^1$ and $A_{32}^2 \tau_{32}^2$, which does not limit in any way the area of application of these formulas, since under all practical conditions $A_{31}^2 \tau_{31}^2 \ll 1$ and $A_{32}^2 \tau_{32}^2 \ll 1$. It must be borne in mind that when $A_{32}^2 \tau_{32}^2 \ll 1$ the quantity $A_{32}^2 \tau_{32} / \Sigma \Lambda_{jk}$ can vary over very wide limits as a result of the great differences in the orders of the quantities $1/\tau_{nm}$ and w_{nm}. Here it must be noted that the expressions for X_2 and $x_2^{\pm}$, X_{23} and $x_{23}^{\pm}$ can be obtained from (5.31) and (5.32), if in each of them we neglect the first equation and in the remaining ones we put $X_{21} = x_{21} = 0$.

We already know that taking into account in the equations for the density matrix the non-diagonal element Z_{21}, which corresponds to the free transition $2 \to 1$, leads to the appearance in the polarization components at the frequency of the signal transition $3 \to 2$ of correction terms of the order of T_2/T_1. These correction terms are important when analyzing the stability of masers and can be omitted when considering the problem of the nonlinearity of masers.

Knowing the solution of (5.11) and (5.32), we can calculate the average value of the polarization of the quantum system being considered. Using (5.34) and (5.35), we will only write the components of the polarization at frequencies close to ω_{32}:

$$m = |\mu_{32}| \operatorname{Re} \left\{ j \frac{X_2}{\gamma_{23}} A_{32} e^{j\omega_{32}t + j\varphi_{32}} + j \left(\frac{X_2}{\gamma_{23} + j\Omega} a + \frac{x_2^+}{\gamma_{23} + j\Omega} A_{32} \right) e^{j(\omega_{32}+\Omega)t + j\varphi} + \right.$$
$$\left. + j \frac{x_2^-}{\gamma_{23}^* - j\Omega} A_{32} e^{j(\omega_{32}-\Omega)t + j2\varphi_{32} - j\varphi} \right\}. \tag{5.36}$$

Therefore, in the polarization of the active material besides components at the frequencies of the strong field (ω_{32}) and the weak field ($\omega_{32} + \Omega$), which are present in the initial field, a combination component at a frequency $\omega_{32} - \Omega$ also appears. In addition, we must draw attention to the appearance in the component at the frequency of the initial weak field of a component proportional to $x_2^+ A_{32}$, which cannot be represented in the form of the product of the amplitude of the weak field and the difference in population. As can be seen from the expression for the difference in population in the zero approximation of X_2 and of the corrections to it x_2^+ and x_2^-, these nonlinear effects become important when the strong field of the signal causes considerable saturation of the transition $3 \to 2$. The appearance of nonlinear components in the polarization and the form of their dependence on the amplitude of the signal field correspond to the results of (62, 63], obtained when calculating the polarization of a two-level quantum system. For saturation of the pumping transition the expressions for X_2 and $x_2^\pm$ are considerably simplified. In this case the dependence of the nonlinear effects considered on the value of the detuning between the frequencies of the strong and weak fields agrees with the dependence obtained in [62] for a two-level system.

If the weak field is nonmonochromatic, in view of the linearity of the approximation considered for the weak field the contribution to the polarization from each spectral component of the weak field does not depend on the remaining components. Thus, for a signal field of the form

$$H_{\text{sig}} = \frac{\hbar}{|\mu_{32}|} A_{32} \cos(\omega_{32}t + \varphi_{32}) + \frac{\hbar}{|\mu_{32}|} \sum_k a_k \cos(\omega_{32}t + \Omega_k t + \varphi_k), \tag{5.37}$$

the polarization equals

$$m = |\mu_{32}| \operatorname{Re} \left\{ j \frac{X_2}{\gamma_{23}} A_{32} e^{j\omega_{32}t + j\varphi_{32}} + j \sum_k \left(\frac{X_2}{\gamma_{23} + j\Omega_k} a_k + \frac{x_{2(k)}^+}{\gamma_{23} + j\Omega_k} A_{32} \right) e^{j(\omega_{32}+\Omega_k)t + j\varphi_k} + \right.$$
$$\left. + j \sum_k \frac{x_{2(k)}^-}{\gamma_{23}^* - j\Omega_k} A_{32} e^{j(\omega_{32}-\Omega_k)t + j2\varphi_{32} - j\varphi_k} \right\}, \tag{5.38}$$

where $x_{2(k)}^+$ and $x_{2(k)}^-$ can be calculated from the formulas for x_2^+ and x_2^-, if we replace a and Ω by the quantities a_k and Ω_k respectively.

Although, with a view to analyzing the modulation distortion, we will investigate in more detail the case of a signal consisting of a strong field at the resonance frequency $\omega_{32} = \omega_{32}^0 \equiv \omega$ and two weak fields with frequencies $\omega \pm \Omega$ symmetrical relative to the frequency ω:

$$H = \frac{\hbar}{|\mu_{32}|} \operatorname{Re} \{ A e^{-j\varphi - j\omega t} + a_1 e^{-j\Phi_1 - j(\omega+\Omega)t} + a e^{j\Phi_2 + j(\omega+\Omega)t} \}. \tag{5.39}$$

We will present in more detail the formula for the polarization produced by a field of this form at frequencies close to the signal frequency:

$$m = |\mu_{32}| \operatorname{Re} \left\{ -j\tau_{23} X_2 A e^{-j\varphi - j\omega t} - j\tau_{32} X_2 \left[\frac{a_1 e^{-j\Phi_1}}{1 - j\Omega\tau_{23}} - BA^2 (a_1 e^{-j\Phi_1} + a_2 e^{-j2\varphi + j\Phi_2}) \right] e^{-j(\omega+\Omega)t} - \right.$$
$$\left. - j\tau_{23} X_2 \left[\frac{a_2 e^{j\Phi_2}}{1 - j\Omega\tau_{23}} - BA^2 (a_2 e^{j\Phi_2} + a_1 e^{j2\varphi - j\Phi_1}) \right] e^{j(\omega-\Omega)} \right\}, \tag{5.40}$$

where the coefficient B which determines the value of the nonlinear effects depends on the field intensity A [see expression (5.35) for x_2^+]:

$$B = \frac{\frac{1}{2}\left[\Lambda_{11}-j\Omega-\frac{\Lambda_{21}}{2}+\frac{3}{8}\left(\frac{1}{\gamma_{13}-j\Omega}+\frac{1}{\gamma_{13}^{*}-j\Omega}\right)A_{31}^{2}\frac{\tau_{23}}{1-j\Omega\tau_{23}}\left(1+\frac{1}{1-j\Omega\tau_{23}}\right)\right]}{F^{*}}. \tag{5.41}$$

When the pumping transition is saturated

$$X_{2} = \frac{-\left(\Lambda_{2}-\frac{\Lambda_{1}}{2}\right)}{\Lambda_{22}-\Lambda_{12}/2+{}^{3}/_{4}A^{2}\tau_{23}}, \tag{5.42}$$

which is equivalent to formula (5.2) for the population inversion of the signal transition taking into account the effect of saturation by the signal field. The coefficient B under these conditions is

$$B = \frac{\frac{3}{8}\frac{\tau_{23}}{1-j\Omega\tau_{23}}\left(1+\frac{1}{1-j\Omega\tau_{23}}\right)}{\Lambda_{22}-j\Omega-\Lambda_{12}/2+{}^{3}/_{4}A^{2}\frac{\tau_{23}}{1-j\Omega\tau_{23}}}. \tag{5.43}$$

Note that if the weak sideband components are equal in amplitude and the phases satisfy the condition $\Phi_1 + \Phi_2 = \pi - 2\varphi$, the correction terms in (5.40) become zero by virtue of the mutual compensation of the symmetrical combination components, and the nonlinear effects mentioned above disappear.

§4. Fields at the Output of a Traveling-Wave Maser

The determination of the fields at the output of a traveling-wave maser, the polarization of the active material of which is given by formula (5.40), requires the specification of the type of oscillations and the field distribution in the slow-wave structure. We will therefore consider an amplifying medium in which a plane wave propagates in the direction x. The dielectric constant of this medium ε takes into account the slowing down of the traveling wave. It is obvious that the nonlinear distortions which occur when plane waves are amplified will be characteristic for traveling-wave masers of arbitrary construction.

From the wave equation

$$\left(\frac{\partial^2}{dx^2}-\frac{\varepsilon}{c^2}\frac{\partial^2}{\partial t^2}\right)H = 4\pi N\frac{\partial^2}{dx^2}m, \tag{5.44}$$

where N is the density of active particles for a field of the form (5.39) and the polarization (5.40) corresponding to it, we can obtain a set of equations for the various spectral components:

$$\left(\frac{d^2}{dx^2}+\varepsilon\frac{\omega^2}{c^2}\right)Ae^{-j\varphi} = -j4\pi N\frac{|\mu_{32}|^2}{\hbar}\tau_{23}\frac{d^2}{dx^2}(X_2 Ae^{-j\varphi}),$$

$$\left(\frac{d^2}{dx^2}+\varepsilon\frac{(\omega+\Omega)^2}{c^2}\right)a_1 e^{-j\Phi_1} = -j4\pi N\frac{|\mu_{32}|^2}{\hbar}\tau_{23}\frac{d^2}{dx^2}\left[\frac{1}{1-j\Omega\tau_{23}}X_2 a_1 e^{-j\Phi_1}-BX_2 A^2(a_1 e^{-j\Phi_1}+a_2 e^{-j2\varphi+j\Phi_2})\right],$$

$$\left(\frac{d^2}{dx^2}+\varepsilon\frac{(\omega-\Omega)^2}{c^2}\right)a_2 e^{j\Phi_2} = j4\pi N\frac{|\mu_{32}|^2}{\hbar}\tau_{23}\frac{d}{dx^2}\left[\frac{1}{1-j\Omega\tau_{23}}X_2 a_2 e^{j\Phi_2}-BX_2 A^2(a_2 e^{j\Phi_2}+a_1 e^{j2\varphi-j\Phi_1})\right]. \tag{5.45}$$

It is convenient here to introduce the parameters α and σ, which characterize respectively the amplifying properties of the inverted levels of the transition $3 \to 2$ and the extent to which they are susceptible to saturation by the strong field A, and also the parameters β, δ, and $\varkappa$, which connect the value of the detuning Ω with the probabilities of relaxation transitions and of transitions induced by the pumping field A_{31}. We will set

$$4\pi N\frac{|\mu_{32}|^2}{\hbar}\tau_{23}X_2\frac{\omega\sqrt{\varepsilon}}{c} = \frac{2\alpha}{1+\sigma^2 A^2}, \quad B = \frac{\delta\sigma^2}{\varkappa^2+\sigma^2 A^2}, \tag{5.46}$$

$$\frac{1}{1-j\Omega\tau_{23}} = \beta, \quad \frac{1}{2}\left(1 + \frac{1}{1-j\Omega\tau_{23}}\right) = \delta, \tag{5.46}$$

where

$$\alpha = 2\pi N \frac{|\mu_{32}|^2}{\hbar} \tau_{23} \frac{\Lambda_{11}\Lambda_2 + \Lambda_{21}\Lambda_1 + \left(\Lambda_2 - \frac{1}{2}\Lambda_1\right)\left(\frac{1}{\gamma_{13}} + \frac{1}{\gamma_{13}^*}\right)\frac{A_{31}^2}{2} \frac{\omega}{c} \sqrt{\varepsilon}}{\Lambda_1\Lambda_{22} - \Lambda_{12}\Lambda_{21} + \left(\Lambda_{22} - \frac{1}{2}\Lambda_{12}\right)\left(\frac{1}{\gamma_{13}} + \frac{1}{\gamma_{13}^*}\right)\frac{A_{31}^2}{2}},$$

$$\sigma^2 = \frac{\Lambda_{11} - \frac{1}{2}\Lambda_{21} + \frac{3}{8}\left(\frac{1}{\gamma_{13}} + \frac{1}{\gamma_{13}^*}\right)A_{31}^2}{\Lambda_{11}\Lambda_{22} - \Lambda_{12}\Lambda_{21} + \left(\Lambda_{22} - \frac{1}{2}\Lambda_{12}\right)\left(\frac{1}{\gamma_{13}} + \frac{1}{\gamma_{13}^*}\right)\frac{1}{2}A_{31}^2} \tau_{23}, \tag{5.47}$$

$$\varkappa = (1 - j\Omega\tau_{23}) \frac{\Lambda_{11} - \frac{1}{2}\Lambda_{21} + \frac{3}{8}\left(\frac{1}{\gamma_{13}-j\Omega} + \frac{1}{\gamma_{13}^*-j\Omega}\right)A_{31}^2}{\Lambda_{11} - j\Omega - \frac{1}{2}\Lambda_{21} + \frac{3}{8}\left(\frac{1}{\gamma_{13}-j\Omega} + \frac{1}{\gamma_{13}^*-j\Omega}\right)A_{31}} \times$$

$$\times \frac{(\Lambda_{11} - j\Omega)(\Lambda_{22} - j\Omega) - \Lambda_{12}\Lambda_{21} + \left(\Lambda_{22} - j\Omega - \frac{1}{2}\Lambda_{12}\right)\left(\frac{1}{\gamma_{13}-j\Omega} + \frac{1}{\gamma_{13}^*-j\Omega}\right)\frac{1}{2}A_{31}^2}{\Lambda_{11}\Lambda_{22} - \Lambda_{12}\Lambda_{21} + \left(\Lambda_{22} + \frac{1}{2}\Lambda_{11}\right)\left(\frac{1}{\gamma_{13}-j\Omega} + \frac{1}{\gamma_{13}^*-j\Omega}\right)\frac{1}{2}A_{31}^2}.$$

When the pumping transition is saturated the expressions for the constants α, σ^2, and $\varkappa$ are considerably simplified

$$\alpha = 2\pi N \frac{|\mu_{32}|^2}{\hbar} \tau_{23} \frac{\Lambda_2 - \frac{1}{2}\Lambda_1}{\Delta_{22} - \frac{1}{2}\Lambda_{12}} \frac{\omega}{c} \sqrt{\varepsilon},$$

$$\sigma^2 = \frac{\frac{3}{4}\tau_{23}}{\Lambda_{22} - \frac{1}{2}\Lambda_{12}}, \tag{5.48}$$

$$\varkappa = (1 - j\Omega\tau_{23})\left(1 - \frac{j\Omega}{\Lambda_{22} - \frac{1}{2}\Lambda_{12}}\right).$$

To solve (5.45) we represent the required functions A_1, a_1, and a_2 in the form of products of rapidly oscillating phase factors and complex amplitudes, where the amplitudes will be expressed in units which determine the degree of saturation:

$$Ae^{-j\varphi} = \frac{1}{\sigma} V(x) e^{j\frac{\omega}{c}\sqrt{\varepsilon}x},$$

$$a_1 e^{-j\Phi_1} = \frac{1}{\sigma} V_1(x) e^{j\frac{\omega+\Omega}{c}\sqrt{\varepsilon}x}, \tag{5.49}$$

$$a_2 e^{j\Phi_2} = \frac{1}{\sigma} V_2(x) e^{-j\frac{\omega-\Omega}{c}\sqrt{\varepsilon}x}.$$

Because of the smallness of the parameter $\left[\frac{\alpha c}{\omega\sqrt{\varepsilon}}\right]$ we can assume that the complex amplitudes V, V_1, and V_2 vary very little over the length $\left[1\big/\frac{\omega\sqrt{\varepsilon}}{c}\right]$, and we obtain from (5.45) for these amplitudes the following approximate equations:

$$\frac{dV}{dx} = \alpha \frac{V}{1+VV^*}, \tag{5.50}$$

$$\frac{dV}{dx} = \alpha\beta \frac{V_1}{1+VV^*} - \alpha\delta \frac{VV^*V_1 + V^2 V_2}{(1+VV^*)(\varkappa+VV^*)},$$
$$\frac{dV}{dx} = \alpha\beta \frac{V_2}{1+VV^*} - \alpha\delta \frac{VV^*V_2 + V^{*2}V_1}{(1+VV^*)(\varkappa+VV^*)}.$$
(5.50)

We see from the equations obtained that the strong field does not depend on the weak fields. Moreover, the presence of a strong field leads to the appearance of coupling between the symmetrical components of the weak field. This coupling causes the appearance of nonlinear effects. We see from the first equation also that only the modulus of the complex amplitude of the strong field V changes in proportion to the amplification. Since by the choice of the beginning of the reading of time we can select any phase for the initial value of V, we will assume the function V to be real. Then, taking into account that $V = V^*$, we can carry out further conversions and obtain equations, which are not connected to one another, for the sum and difference of the weak fields $V_1 + V_2$ and $V_1 - V_2$:

$$\frac{dV}{dX} = \alpha \frac{V}{1+V^2},$$
$$\frac{d(V_1+V_2)}{dX} = \alpha\beta \frac{V_1+V_2}{1+V^2} - 2\alpha\delta \frac{V^2}{(1+V^2)(\varkappa+V^2)}(V_1+V_2),$$
$$\frac{d(V_1-V_2)}{dX} = \alpha\beta \frac{V_1-V_2}{1+V^2}.$$
(5.51)

These equations are easily integrated, which enables us to relate the input and output values of the amplitudes V, V_1, and V_2:

$$\ln \frac{V_{\text{out}}^2}{V_{\text{in}}^2} + V_{\text{out}}^2 - V_{\text{in}}^2 = 2\alpha l,$$
(5.52)

$$\frac{(V_1+V_2)_{\text{out}}}{(V_1+V_2)_{\text{in}}} = \left(\frac{V_{\text{out}}}{V_{\text{in}}}\right)^\beta \frac{(\varkappa+V_{\text{in}}^2)^\delta}{(\varkappa+V_{\text{out}}^2)^\delta},$$
(5.53)

$$\frac{(V_1-V_2)_{\text{out}}}{(V_1-V_2)_{\text{in}}} = \left(\frac{V_{\text{out}}}{V_{\text{in}}}\right)^\beta,$$
(5.54)

where l is the length of the maser, and the subscripts "in" and "out" relate to the input and output values respectively.

Formula (5.52) describes the reduction in the gain of the central component when the maser is saturated and corresponds to the results obtained in § 6, Chapter VII. We see from formula (5.53) that if at the input the sideband components satisfy the condition $V_1 + V_2 = 0$ (which corresponds to a frequency-modulated input signal), this condition also remains true at the output (the signal remains frequency-modulated). Similarly, we see from (5.54) that an amplitude-modulated signal ($V_1 - V_2 = 0$) after amplification remains amplitude-modulated. If $(V_1 + V_2)_{\text{in}} \neq 0$ and $(V_1 - V_2)_{\text{in}} \neq 0$, the input relation between the quantities V_1 and V_2 is destroyed. In particular, if at the input there is a signal in one sideband ($V_{1\text{in}} \neq 0$, $V_{2\text{in}} = 0$), a field of the combination frequency $V_{2\text{out}}$ will appear at the output.

We will compare the cases of frequency and amplitude modulation, in which the amplification of the sideband components is described by the formulas (5.54) and (5.53) respectively. For frequency modulation the sideband components are amplified in accordance with the frequency characteristic of the linear maser, the gain of which at the central frequency G is determined by the relation (5.54). In this case there are no nonlinear distortions of the frequency-modulated signal.

With amplitude modulation the sideband components are changed in a very complex way. For certain parameters of the maser and the degree of saturation at the input the sideband components may not be amplified but absorbed. These peculiarities of the propagation of an amplitude-modulated signal, described by the factor $[(\varkappa + V_{in}^2)/(\varkappa + V_{out}^2)]^\delta$ in formula (5.53), can lead to considerable distortion.

Note that in the case when the weak field at the input contains not two symmetrical components but has a very complex spectral composition, formulas (5.53) and (5.54) remain true provided that we mean by V_1 and V_2 any two components with symmetrical frequencies.

§ 5. The Paramagnetic Maser

In the case of paramagnetic masers saturation of the pumping transition $A_{31}^2/|\Lambda_{mn}\gamma_3| \gg 1$ is carried out. In addition, for paramagnetic masers we can rewrite the expressions for the constants α, σ^2, $\varkappa$, and β, introducing into our consideration the spin-spin relaxation time T_2 and the spin-lattice relaxation time T_1:

$$T_2 = \tau_{23} \text{ and } T_1 = \frac{1}{\Lambda_{22} - \frac{1}{2}\Lambda_{12}} = \frac{2}{2w_{21} + w_{12} + w_{32} + 2w_{23}}. \tag{5.55}$$

From expressions (5.48) and (5.46) taking into account (5.55), we obtain

$$\alpha = 2\pi N \frac{|\mu_{32}|^2}{\hbar} T_2 T_1 (w_{21} - w_{12} - w_{32} + w_{23}),$$

$$\sigma^2 = \frac{3}{4} T_2 T_1,$$

$$\varkappa = (1 - j\Omega T_2)(1 - j\Omega T_1), \tag{5.56}$$

$$\beta = 1/(1 - j\Omega T_2),$$

$$\delta = 1/2 + 1/2/(1 - j\Omega T_2).$$

These values of the constants enable us to represent formulas (5.52)-(5.54) more clearly. We will first introduce the power gain for the signal field of the resonance frequency $G = V_{out}^2/V_{in}^2$. With this formula (5.52) is reduced to the form (5.13)

$$G = G_0 e^{-V_{in}^2 (G-1)}, \tag{5.57}$$

where G_0 is the gain in the absence of saturation. The amplitudes of the sideband components can then be expressed in terms of the gain G for the central component

$$\frac{(V_1 + V_2)_{out}}{(V_1 + V_2)_{in}} = G^{\frac{1/2}{1-j\Omega T_2}} \left[\frac{(1 - j\Omega T_2)(1 - \Omega T_1) + V_{in}^2}{(1 - j\Omega T_2)(1 - j\Omega T_1) + GV_{in}^2} \right]^{\frac{1}{2} + \frac{1/2}{1-j\Omega T_2}}, \tag{5.58}$$

$$\frac{(V_1 - V_2)_{out}}{(V_1 - V_2)_{in}} = G^{\frac{1/2}{1-j\Omega T_2}}. \tag{5.59}$$

These formulas give in explicit form the dependence of the amplitudes of the sideband components at the output of the saturated maser on the modulation frequency Ω, and enable us to determine the values of such distortions as the reduction in the modulation depth, the intensity of the combination frequency fields, and the change in the phase shift of the sideband components, when the maser is saturated (see §6, Chapter V).

The input signal of the form (5.39) describes the spectral expansion of a frequency-modulated signal with a small modulation index for the conditions $V_{in} = -V_{2\,in}$. It follows from (5.58) and (5.59) that when masers are saturated, as was pointed out in the previous section, there is no distortion of a frequency-modulated signal. For large modulation indexes the signal (5.39) must be supplemented by components at frequencies $\omega \pm n\Omega$, the amplitudes of which are mutually antisymmetric, as a result of which in this case there are no nonlinear distortions.

With amplitude modulation $V_{1\text{in}} = V_{2\text{in}}$, and formula (5.58) gives the gain of the field of the sideband component $g^{1/2} = \dfrac{V_{1\text{out}}}{V_{1\text{in}}} = \dfrac{V_{2\text{out}}}{V_{1\text{in}}} = \dfrac{(V_1 + V_2)_{\text{out}}}{(V_1 + V_2)_{\text{in}}}$. We see from the formula that the sideband components of an amplitude-modulated signal are amplified to a lesser extent than the field of the carrier frequency. The first factor in (5.58) describes the reduction in the gain similar to linear distortion of the modulation depth. In our case these distortions are complicated by the dependence of the resonance gain on the intensity of the carrier field. The frequency dependence of these distortions are determined by the time T_2.

The second factor is due to the interaction of the weak fields in the presence of the strong field. Its frequency dependence is determined by the times T_1 and T_2. The role of this factor becomes important when the intensity of the strong field becomes comparable with $|(1 - j\Omega T_2) \times (1 - j\Omega T_1)| = \sqrt{(1 - \Omega^2 T_2 T_1)^2 + \Omega(T_2 + T_1)^2}$. As can be seen from (5.58), when $g > 1$, we always have $|g| < G$, i.e, for amplification in an amplifier which is partly saturated a reduction in the modulation depth occurs.

From formula (5.58) it is easy to obtain the two limiting cases in which there is no nonlinear distortion. These are the obvious cases of very small saturation (the maser is linear) and the completely saturated maser (the medium is transparent). The greatest deviations from linearity occur when $\Omega T_1 < 1$. Therefore, the estimates given in [57] cannot be used to determine the degree of distortion of the modulation depth for practical modulation frequencies exceeding $1/T_1$.

For $\Omega \approx 1/T_2$ and for practically possible intensities of the strong field $V^2 \ll 1/T_2^2$ the nonlinear distortions due to the second factor in (5.58) become unimportant. The distortions governed by the first factor will be less than under the linear operating conditions due to the widening of the bandwidth as a result of the saturation of the resonance amplification.

Note that for amplitude modulation with certain modulation frequencies not only does the modulation depth change for each of the frequencies but also the relations between the amplitudes of the sidebands are distorted due to the dependence of the effects considered on the value of the detuning Ω.

Therefore, the analysis carried out for the case of a strong field at the resonance frequency which is associated with the sidebands enables us to determine the nonlinear distortions in traveling-wave masers.

To conclude this section we will give expressions for the complex amplitudes of the sideband components:

$$V_{1\text{out}} = \tfrac{1}{2} G^{\frac{1/2}{1-j\Omega T_2}} \left[V_{1\text{in}} - V_{2\text{in}} + (V_{1\text{in}} + V_{2\text{in}}) \left(\frac{(1-j\Omega T_2)(1-j\Omega T_1) + V_{\text{in}}^2}{(1-j\Omega T_2)(1-j\Omega T_1) + GV_{\text{in}}^2} \right)^{1/2 + \frac{1/2}{1-j\Omega T_2}} \right],$$

$$V_{2\text{out}} = \tfrac{1}{2} G^{\frac{1/2}{1-j\Omega T_2}} \left[V_{2\text{in}} - V_{1\text{in}} + (V_{2\text{in}} + V_{1\text{in}}) \left(\frac{(1-j\Omega T_2)(1-j\Omega T_1) + V_{\text{in}}^2}{(1-j\Omega T_2)(1-j\Omega T_1) + GV_{\text{in}}^2} \right)^{1/2 + \frac{1/2}{1-j\Omega T_2}} \right].$$

(5.60)

When $V_{2\text{in}} = 0$ we can conditionally assume that the expression for $V_{1\text{out}}$ gives the amplitude and phase characteristics of the maser for a weak signal in one sideband for a given level of saturation by the strong signal of the resonance frequency. This expression can be used, in particular, to determine the phase distortions of the signal in one sideband when the level of saturation is changed. The expression for $V_{2\text{out}}$ gives in this case the amplitude and phase of the field of the combination frequency.

§6. Analysis of Some Nonlinear Distortions in Traveling-Wave Masers

The expressions obtained in the previous section for the amplitudes of the fields at the output of a traveling-wave maser with saturation enable us, as already pointed out, to carry out an analysis of the nonlinear distortions in any concrete case. In this section we determine the value of those nonlinear distortions which arise when the maser is saturated by the field of the resonance frequency, such as the generation of a combination frequency field when only a signal in one sideband is present at the input, the distortion of the phase of the signal of one sideband, the distortion of the amplitude-modulation depth, and the distortion of the relation between the amplitudes of the various spectral components.

We will rewrite formulas (5.60) for the amplitudes of the fields V_1 and V_2 taking into account the fact that $T_2 \ll T_1$:

$$V_{1\text{out}} = \tfrac{1}{2} G^{\frac{1+j\Omega T_2}{2+2\Omega^2 T_2^2}} \left[V_{1\text{in}} - V_{2\text{in}} + (V_{1\text{in}} + V_{2\text{in}}) \left(\frac{1 - \Omega^2 T_2 T_1 + V_{\text{in}}^2 - j\Omega T_1}{1 - \Omega^2 T_2 T_1 + G V_{\text{in}}^2 - j\Omega T_1} \right)^{\frac{2+\Omega^2 T_2^2 + j\Omega T_2}{2+2\Omega^2 T_2^2}} \right],$$

$$V_{2\text{out}} = \tfrac{1}{2} G^{\frac{1+j\Omega T_2}{2+2\Omega^2 T_2^2}} \left[V_{2\text{in}} - V_{1\text{in}} + (V_{2\text{in}} + V_{1\text{in}}) \left(\frac{1 - \Omega^2 T_2 T_1 + V_{\text{in}}^2 - j\Omega T_1}{1 - \Omega^2 T_2 T_1 + G V_{\text{in}}^2 - j\Omega T_1} \right)^{\frac{2+\Omega^2 T_2^2 + j\Omega T_2}{2+2\Omega^2 T_2^2}} \right]. \quad (5.61)$$

We will consider the amplitude of the wave which appears at the output of the maser at the combination frequency $\omega - \Omega$ when radiation at the frequencies ω (the strong field) and $\omega + \Omega$ (the weak field) is present at the input. Therefore we will assume that $V_{2\text{in}} = 0$. As a measure of the nonlinearity we will take the modulus of the ratio of the output amplitudes $|V_{2\text{out}}/V_{1\text{out}}|$.

In the particular case of very small detuning, i.e., when $\Omega T_1 \ll 1$, it is easy to obtain that

$$\frac{V_{2\text{out}}}{V_{1\text{out}}} = -\frac{(G-1) V_{\text{in}}^2}{2 + (G+1) V_{\text{in}}^2}. \quad (5.62)$$

It is interesting that the spectrum at the output becomes similar to the spectrum of frequency modulation; the amplitudes of the fields at the frequencies $\omega + \Omega$ have opposite signs. It is convenient to write the expression obtained in another form by expressing V_{in}^2 in terms of G and G_0:

$$\frac{V_{2\text{out}}}{V_{1\text{out}}} = -\frac{\ln G_0/G}{2 + \frac{G-1}{G+1} \ln G_0/G}. \quad (5.63)$$

This quantity is always less than unity.

In the more interesting case when the frequency difference is not small in comparison with the rate of population relaxation, an additional phase shift occurs between the output amplitudes. For frequencies $\Omega \ll 1/T_2$

$$\frac{V_{2\text{out}}}{V_{1\text{out}}} = -\frac{(G-1) V_{\text{in}}^2}{2 + (G+1) V_{\text{in}}^2 - 2j\Omega T_1}. \quad (5.64)$$

In this case the intensity of the effect falls:

$$\left| \frac{V_{2\text{out}}}{V_{1\text{out}}} \right| = \frac{\ln G_0/G}{\sqrt{\left[2 + \frac{G+1}{G-1} \ln G_0/G\right]^2 + 4\Omega^2 T_1^2}}. \quad (5.65)$$

For large detuning frequencies $\Omega \gg 1/T_1$, by means of (5.61) we can obtain the approximate expression

$$\left|\frac{V_{2\text{out}}}{V_{1\text{out}}}\right| = \frac{1}{2}\frac{1}{\Omega T_1}\frac{\ln G_0/G}{(1+\Omega^2 T_2^2)^2}, \qquad (5.66)$$

from which we see that when $\Omega \sim 1/T_2$ the nonlinear effect being considered is of the order of magnitude of T_2/T_1, i.e., is very small.

It should be noted that this expression does not hold when $G \to 1$, when $V_{2\text{out}} \to 0$ irrespective of the dependence on Ω.

Therefore, for three different ranges of detuning frequencies three simple formulas are obtained which enable us to determine the relative amplitude of the combination current. The form of the dependence on the initial gain G_0 is common to these formulas, and also the fact that the value of the effect is determined by the degree of saturation of the gain G. The logarithmic dependence of these effects on the gains G_0 and G is characteristic of traveling-wave masers and often manifests itself when analyzing the properties of these masers (the bandwidth, the stability, and the nonlinear distortions).

The value of the nonlinear effect strongly depends on the detuning frequency. For very small detunings the effect is quite large. However, for $\Omega = 10/T_1$ under extreme conditions the field of the combination frequency is 12 dB weaker than the initial field. For $T_1 = 10^{-3}$ sec (rutile) this represents a frequency of 1.5 kHz. As the frequency is increased a further reduction in this effect occurs. For frequencies $\Omega T_1 \gg 1$ it is necessary to use formula (5.66) from which it follows that, for example, for $\Omega T_2 = 1$ and for saturation of the resonance amplification from 30 to 10 dB the relative amplitude of the combination field is 10^{-5}–10^{-7}, i.e., -100 to -140 dB, depending on the value of T_1 of the working material. In formulas (5.65) and (5.66) and in the further discussion we will use for the characteristic of the degree of saturation of the maser the value of its gain at the resonance frequency G. The relation between the quantities G_0, V_{in}, and G is given by formula (5.57). Figure 10 shows the form of this dependence.

The distortions of the phase of the weak signal can be important in a number of phase-measuring and radio-interference circuits. We will consider the phase distortion for the case of a signal in one sideband ($V_2 = 0$). First of all, we note that in all linear resonator systems a phase shift always appears when they are off-tuned from the center frequency. Therefore, we must expect that with saturation of the maser by the field of the resonance frequency the phase shift for the sideband signal will be changed in view of the change (the relative broadening) which occurs in this case in the amplifier frequency characteristic. Essentially this change is a linear phase shift. But besides this strictly linear phase distortions must exist. Thus for small off-tunings, when $\Omega \ll 1/T_2$, we can obtain from (5.61) that the phase of the output field $V_{1\text{out}}$ is determined by the formula

$$\tan\varphi = \frac{-\Omega T_1 \ln G_0/G}{\left(1 + \frac{G}{G-1}\ln G_0/G\right)\left(2 + \frac{G+1}{G-1}\ln\frac{G_0}{G}\right) + 2\Omega^2 T_1^2}. \qquad (5.67)$$

This is important since it indicates the presence of phase shifts due to saturation for detunings for which the deviation of the frequency characteristic from a uniform one is still insignificant.

For frequencies $\Omega \gg 1/T_1$ we obtain from (5.61) that the phase of the field $V_{1\text{out}}$ is equal to

$$\varphi = \frac{\Omega T_2}{2}\frac{\ln G}{1+\Omega^2 T_2^2} - \frac{1}{\Omega T_1}\frac{\ln G_0/G}{(1+\Omega^2 T_2^2)^2}. \qquad (5.68)$$

In this case the linear phase shift is

$$\varphi_0 = \frac{\Omega T_2}{2}\frac{\ln G_0}{1+\Omega^2 T_2^2}. \qquad (5.69)$$

If we subtract this linear phase shift from the expression for φ, we obtain an expression for the change in the phase shift $\Delta\varphi = \varphi - \varphi_0$, due to saturation of the maser. When $\Omega \gg 1/T_1$ this change is

$$\Delta\varphi = -\left(1 + \frac{2/T_2 T_1 \Omega^2}{1 + \Omega^2 T_2^2}\right) \frac{\Omega T_2}{2} \frac{\ln G_0/G}{1 + \Omega^2 T_2^2} \tag{5.70}$$

and if the gain is measured in decibels, it increases linearly as the gain with saturation is reduced.

In the region of frequencies $\Omega \ll 1/T_2$ the phase distortions can be quite considerable for very small frequencies $\Omega \approx 1/T_1$. But when $\Omega = 50/T_1$ the maximum possible change in the phase shift is 0.06 rad when $G_0 = 1000$. When Ω is increased the maximum is shifted to the region of small gains (greater saturation) and for a certain value of Ω begins to fall with $1/\Omega$.

For large detunings, $\Omega \approx 1/T_2$, "pseudolinear" phase distortions are important. For $\Omega T_2 = 0.1$, i.e., within the limits of the practically flat top of the frequency characteristic of the unsaturated maser, the change in phase shift for a complete saturation of the maser is 0.35 rad.

We note again that when $1/T_1 \ll \Omega \approx 1/T_2$ the phase distortion is basically determined by the widening of the maser bandwidth due to saturation of the resonance amplification, i.e., the phase distortion leads to a reduction in the linear phase shift $\varphi_0 = \Omega T_2/(1 + \Omega^2 T_2^2)$. Therefore, $\Delta\varphi$ does not fall to zero as $G \to 1$. The correction due to purely nonlinear effects is of the order of magnitude of T_2/T_1 and is very small.

For amplitude modulation $V_{1in} = V_{2in}$ and $V_{1out} = V_{2out}$. The modulation depth can be characterized by the ratio of the amplitude of the sideband to the amplitude of the carrier. Then the relative change in the modulation depth at the output compared with the input can serve as a measure of the distortion of the modulation. We will introduce the coefficient

$$m = \frac{V_{1\,out}}{V_{out}} \bigg/ \frac{V_{1\,in}}{V_{in}}. \tag{5.71}$$

When there is no distortion $m = 1$, and when the modulation disappears at the output $m = 0$.

In the simplest case $\Omega \ll 1/T_1$

$$m = \frac{1 + \frac{1}{G-1} \ln G_0/G}{1 + \frac{G}{G-1} \ln G_0/G}, \tag{5.72}$$

which is always less than unity. This formula agrees with the results obtained in [57], which were derived for quasi-static modulation. For modulation frequencies which are not too large $\Omega \ll 1/T_2$

$$m = \frac{1 + V_{in}^2 - j\Omega T_1}{1 + G V_{in}^2 - j\Omega T_1} \tag{5.73}$$

and it is necessary to take into account the appearance of a phase shift between the sideband components and the carrier. We will characterize the change in the modulation depth by the modulus $|m|$, which in the case considered equals

$$|m| = \sqrt{\frac{\left(1 + \frac{1}{G-1} \ln G_0/G\right)^2 + \Omega^2 T_1^2}{\left(1 + \frac{G}{G-1} \ln G_0/G\right)^2 + \Omega^2 T_1^2}}. \tag{5.74}$$

For the same saturation in this case the modulation depth is reduced less than in the previous case of very slow modulation frequencies, which give rise to a quasi-static change in the gain.

In the case of large modulation frequencies $\Omega \gg 1/T_1$

$$|m| = \left(1 - \frac{T_2}{2T_1}(3 + \Omega^2 T_2^2)\frac{\ln G_0/G}{(1+\Omega^2 T_2^2)^2}\right) e^{-\frac{1}{2}\frac{\Omega^2 T_2^2}{1+\Omega^2 T_2^2}\ln G} . \tag{5.75}$$

If now, as was done in the case of phase distortions, we distinguish the linear reduction in the modulation depth

$$|m_0| = e^{-\frac{1}{2}\frac{\Omega^2 T_2^2}{1+\Omega^2 T_2^2}\ln G_0}, \tag{5.76}$$

we can obtain the expression

$$\left|\frac{m}{m_0}\right| = \left(1 - \frac{1}{2}\frac{T_2}{T_1}(3+\Omega^2 T_2^2)\frac{\ln G_0/G}{(1+\Omega^2 T_2^2)^2}\right) e^{\frac{1}{2}\frac{\Omega^2 T_2^2}{1+\Omega^2 T_2^2}\ln G_0/G}, \tag{5.77}$$

which characterizes the value of the change in the modulation depth when the maser is saturated. As in the case of phase distortions, this change is mainly determined by the degree of broadening of the bandwidth of the maser due to saturation of the resonance amplification. The correction due to nonlinear effects is small. For small modulation frequencies the nonlinear effects are decisive.

For a maser with an initial gain G_0 = 1000 the depth of slow modulation is reduced by a factor of 4 for a drop in gain from 30 to 15 dB. When the modulation frequency is increased the drop in the modulation depth with saturation is reduced. For $\Omega = 10/T_1$ the maximum reduction in the modulation depth is 25%. For a further increase in Ω the reduction in the modulation disappears and for $\Omega \approx 1/T_2$ when the maser is saturated the modulation depth increases due to the reduction in the linear distortion of the modulation.

The nonlinear distortions in masers have an unusual form. As distinct from the usual radio engineering devices, where a reduction in the gain due to saturation is due to the strong nonlinear distortions of the spectrum of the amplified signals, in masers even considerable saturation does not lead to any serious distortions of the spectrum for frequencies $\Omega \gg 1/T_1$, i.e., practically over the whole bandwidth where, as we saw, purely nonlinear effects have an order of magnitude $1/\Omega T_1$, which enables them to be neglected in a number of practically important cases.

Consider the normalized amplitude frequency characteristics $K(\Omega) = (1/\sqrt{G_0})|V_{1out}/V_{1in}|$ for $V_{2in} = 0$ shown in Fig. 13. The curves were plotted for G = 500 (27 dB), G = 100 (20 dB), and G = 31.6 (15 dB) for an initial gain G_0 = 1000 (30 dB).

In order to show more clearly the details of the frequency dependence for small $\Omega \approx 1/T_1$, we chose a rather artificial abscissa axis scale, the steps being linear from $\Omega = 0$ to $\Omega = \pm 20/T_1$. Then the frequency variation is interrupted, and the graph begins from the value $\Omega = \pm 0.01/T_2$

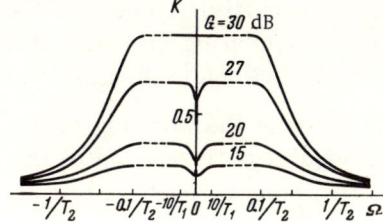

Fig. 13. Normalized frequency characteristics.

and is continued further linearly with frequency, but in steps which strongly decrease. The discontinuity at the zero is due to the effect of the combination term $V_{2\text{out}}$ for $\Omega = 0$. We see from the figure that in the frequency range $10/T_1 < |\Omega| < 0.1/T_2$ the frequency characteristic of the traveling-wave maser with a nominal gain of 30 dB is practically uniform for any saturation level, which illustrates well the relations obtained above.

Therefore, in many practically important cases of the use of masers in communication lines, we can assume that the saturation of the maser reduces the information capacity of the communication line only to the extent of reducing the sensitivity and does not lead to the appearance of spurious signals or nonlinear distortion of the useful signals.

The high linearity of masers is due to the presence in their working material of two relaxation processes: one, which operates with a characteristic time T_1 and is responsible for the restoration of equilibrium with the radiation over the whole system of energy levels, and the other which operates with a characteristic time T_2 and is responsible for the restoration of equilibrium inside a single level. For this reason changes in phase due to fluctuations in the frequency and intensity of the pumping radiation, as we saw, are also quantities of the order of T_2/T_1.

Therefore, masers successfully combine high sensitivity, stability, and small nonlinear distortions.

CHAPTER VI

Experimental Investigations

In the previous chapters we developed the theory of masers and determined their chief properties. This chapter contains the results of specially made experimental investigations on masers [41, 64-71] and gives a description of the technical improvements which are useful when making practical experiments [72-74] in quantum electronics. In this chapter we also describe the results obtained and published by R. M. Martirosyan, R. L. Sorochenko, and A. M. Prokhorov [53], and Yu. P. Pimenov and A. M. Prokhorov [5], which are closely related to the theoretical investigations described in the previous chapters. In addition, we will also describe the results obtained by B. B. Krynetskii, G. P. Kuz'min, and A. E. Shirkov [76], which are directly related to our investigation.

§1. The Traveling-Wave Maser

For many applications including radio astronomy, the design of traveling-wave masers in the three-centimeter band is of great interest. Although in this band the bandwidth of resonator masers depends to a considerable extent on the width of the signal transition line, nevertheless, without a considerable increase in the number of stages it is not possible to approach significantly to the bandwidth of a traveling-wave maser, particularly for high gains. In addition, traveling-wave masers are considerably more stable than resonator masers and, what is particularly important for attaining maximum gain in sensitivity, because the nonreciprocal elements in them are cooled to the temperature of the operating material, their effective noise is less than the noise of resonator masers with circulators which are at room temperature.

In the maser described further we used pink ruby as the working substance. In the three-centimeter band the most advantageous method of population inversion of the signal transition is the so-called symmetrical or push-pull method [77]. If the intensity vector of the external constant magnetic field makes an angle of 54°45' with the direction of the axis of trigonal symmetry

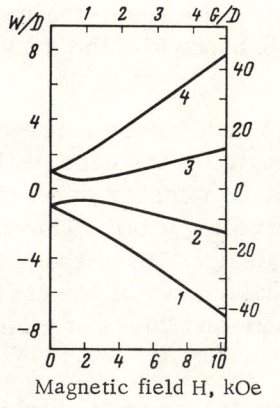

Fig. 14. Symmetrical energy-level diagram in ruby

of the crystal the splitting of the energy levels is symmetrical in relation to the abscissa axis. Then, choosing as the signal transition 3 → 2 (Fig. 14), we can, using pumping radiation of the same frequency $\nu_{31} = \nu_{42}$, simultaneously enrich the upper level of the signal transition (pumping transition 1 → 3) and deplete its lower level (pumping transition 2 → 4). In this case the pumping radiation of wavelength 1.27 cm and the magnetic field of 4300 Oe correspond to a signal of wavelength 3.2 cm. The accuracy of orientation of the constant magnetic field in relation to the crystal axis must not be worse than ± 15'.

The squares of the moduli of the matrix element components of the transition probability at the signal frequency ν_{32} in units of $g\beta$ are $|\mu_{32}|_x = 0.036$, $|\mu_{32}|_y = 0.76$, and $|\mu_{32}|_z = 0.73$, where the z axis is parallel to the trigonal axis of the crystal and the magnetic field is in the OXZ plane. We draw attention to the high circularity of these components which enables us, when there is a high-frequency magnetic field with circular polarization present, to realize unidirectional amplification.

We use a post system as the basis for the design of the slow-wave structure [78]. Periodic post structures possess the properties of slowing electromagnetic waves. Moreover, since each post is similar to a grounded quarter-wave dipole, at the base of the post where the high-frequency magnetic field is concentrated there must exist a region with circular polarization of this field. As distinct from the published constructions of traveling-wave masers in which coaxial excitation of the post system was used, we used waveguide excitation, which is more convenient for the three-centimeter band. The junction of the waveguide excitation is shown schematically in Fig. 15. The height of the posts at the piece of waveguide is of the order of 3 cm and gradually reduces to zero. The side walls of the case of the system form a pyramidal horn of such dimensions that the critical wavelength in each perpendicular cross section of the system, considered as a pi-shaped waveguide, was larger than the critical wavelength of the slow-wave structure. Such a method of high-frequency matching ensures a smooth transformation of the high-frequency field of the conducting waveguide (the TE_{10} mode in rectangular waveguide) to the field of the slow-wave structure.

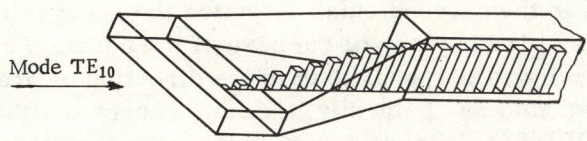

Fig. 15. Waveguide excitation of the post system.

For a reduction of the dimensions of the system as a whole the height of the conducting waveguide was reduced from 10 to 4 mm. Over a 5% bandwidth the VSWR of the transition did not exceed 1.5.

To use the slow-wave structure effectively it is necessary that the energy density of the traveling wave should greatly exceed the energy density in the case of the usual waveguide propagation. In the previous description this increase in energy density was characterized by the value of the delay of the group velocity of the traveling wave. The well-known method of dispersion characteristics was used in the original design [79] of the post structure. However, this method, as also the method of measuring the phase delay of the envelope of the modulated traveling wave [80], requires laborious measurements and does not give a clear picture of the propagation of the fields in the structure.

A direct method of determining the degree of increase of the high-frequency energy density of the traveling wave in the slow-wave structure as applied to the design of a paramagnetic traveling-wave maser is to use electron paramagnetic resonance in the free radical α-α-diphenyl-β-β-picrylhydrazyl (DPPH). It is obvious that the ratio of the values of the EPR lines for one and the same specimen of DPPH, placed in the slow-wave structure or in the conducting waveguide, is determined by the ratio of the corresponding densities of the high-frequency magnetic energy. In this case, when the high-frequency magnetic fields have the same distribution in the conducting feeder and in the system under investigation the ratio of the values of the EPR lines unambiguously define the ratio of the group velocities. Then, for square-law detection the value of the EPR lines is inversely proportional to the corresponding group velocity.

The distribution of the magnetic fields both in the system under investigation and in the conducting feeder can be easily determined experimentally by displacing the specimen of DPPH and by measuring the value of the EPR lines. Note that a knowledge of the magnetic field distribution can help to clarify the electric field distribution.

Besides determining the degree of concentration and the distribution of the electromagnetic energy in the slow-wave structure, the use of DPPH enables us to clarify the configuration of the magnetic fields. For a magnetized specimen of DPPH the unpaired electrons precess in the plane perpendicular to the magnetizing field. Therefore, the intensity of the EPR lines is determined by the value of the high-frequency magnetic field with circular polarization of the corresponding direction which is in this plane. This fact enables us, by changing the orientation and the polarity of the magnetizing field, to carry out a complete analysis of the polarization of the high-frequency magnetic fields of the system under investigation.

Using this method we investigated slow-wave structures constructed when designing the maser. The DPPH powder was placed in a thin-walled polystyrene tube 50 mm long, and of internal diameter 1 mm. It was found that the energy density increased in comparison with a standard waveguide. This increase depends on the geometry of the posts. The best results were obtained with a comb with rectangular posts of height 7 mm with a cross section of 0.5 × 0.6 mm and a spacing of 1 mm. In this case the energy density of the traveling wave was increased by a factor of 55. The energy of the high-frequency magnetic field was concentrated mainly at the base of the posts. At a distance of 2 mm from the base of the posts both in the direction of the posts and in the perpendicular direction the energy density fell to half. This field is elliptically polarized in the plane of the base of the posts. For the geometry mentioned the major axis of the ellipse was perpendicular to the direction of propagation of the traveling wave. The eccentricity equals 1.5. Naturally, the field energy distribution was measured by means of a specimen of DPPH the dimensions of which were less than the width of a post.

The eccentricity of 1.5 indicates that the wave is close to a circularly polarized wave. The energy density in a wave with circular polarization of some definite sign exceeds by a factor of 25 the energy density in a wave with circular polarization of the opposite sign. Since the directions of rotation of the magnetic field vector are opposed on the left and right of the posts of the comb, in some one specific sign of the circular polarization on one side of the comb the energy density of the forward wave exceeds by a factor of 25 the density of the backward wave, whereas on the other side of the comb the reverse relation exists.

The presence of regions of the high-frequency magnetic field with circular polarization of different signs for the forward and backward waves, and also the high circularity of the matrix elements of the signal transition probability are used to obtain nonreciprocal, unidirectional amplification and attenuation. This can be realized if a crystal in the form of a long rod grown perpendicular to the trigonal axis is attached to the slow-wave structure so that its trigonal axis lies in the plane of the base of the comb system, perpendicular to the rods (Fig. 16).

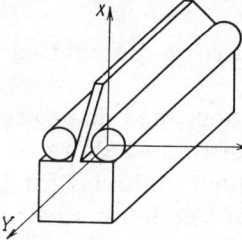

Fig. 16. Orientation of the crystals on the slow-wave structure.

Since the magnetic field and not the electric field is concentrated at the base of the rods the presence in this part of the slow-wave structure of ruby crystals does not appreciably alter the frequency band of the system and does not greatly increase the slowing of the group velocity. The ruby crystals for this maser were specially grown in the Institute of Crystallography of the Academy of Sciences of the USSR in the form of thin circular rods of length 100-110 mm. We specially draw attention to the fact that in each rod, the trigonal axis was perpendicular to its length and lay very accurately in one plane over the whole of its length. Before placement in the slow-wave structure the shape of the crystals was reduced to a cylindrical diameter of 2 mm. The working length was 100 mm. The fact that a single crystal of high uniformity filled the whole working length of the traveling-wave system considerably simplified the tuning and start-up of the maser.

The crystals were attached to the comb with cotton threads. This method turned out to be better than various adhesives and better than soldering the partially silver-plated rods. When attached by the threads the orientation of the crystals was excellently maintained despite many coolings-down to liquid-helium temperatures (Fig. 17). The amplifying crystal (the density of Cr^{3+} ions was of the order of 0.07% in the original mixture) was attached on one side of the rods of the slow-wave structure, and the isolating crystal (density 2%) on the other side.

Great attention was paid to improving the quality of the contact between the base of the comb and the case of the slow-wave structure. The presence of longitudinal slots between these parts led to the appearance of appreciable losses. When the system was cooled-down to liquid

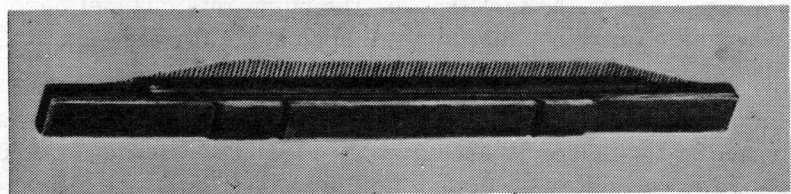

Fig. 17. Attachment of the crystals to the slow-wave structure.

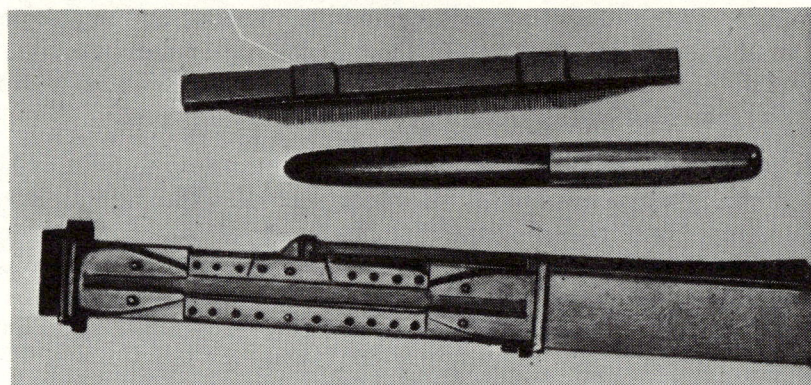

Fig. 18. The slow-wave structure in dismantled form.

helium temperatures the losses of this type often increased due to temperature deformations. The contact was considerably improved by means of thin washers of metallic indium between the strongly compressed case and the comb. The best material for the slow-wave structure is high-frequency copper. In this case manufacture of the comb on a milling machine gave better results than the use of an electron beam.

The ohmic loss in the system was 8-10 dB, and the nonreciprocal reverse loss due to the isolating action of the pink ruby was 25 dB.

At a temperature of 4.2°K we obtained a net amplification of 10 dB, and with pumping of the helium down to 1.8°K a net amplification of 21 dB was obtained, and in this case the bandwidth exceeded 20 MHz. These results considerably exceed those obtained in [81] mainly because we used completely uniform ruby rods, and not a collection of crystals as used in [81].

Measurements of the reverse losses in red ruby with the pumping radiation and without it, and also the direct paramagnetic losses (with the pumping radiation the direct paramagnetic losses are negative) enabled us to determine such characteristics of the material as the intensity of the absorption line of red (28 dB) and pink (10 dB) ruby, the inversion of the absorption line of red ruby (3.4), the degree of nonreciprocity in the interaction of the backward wave with the pink ruby, or, which is the same thing, of the forward wave with the red ruby (0.09). The data given in the brackets relate to a temperature of 1.8°K.

When using the waveguide method of exciting the slow-wave structure at the signal frequency, the pumping radiation in the maser is fed into the case of the slow-wave structure through an opening in the middle of its side wall (Fig. 18) and not through the end of the system, as is usual. In this case the pumping field is introduced by the rectangular waveguide so that the E vector of this field is perpendicular to the lateral plane of the rods of the system. The pump feeds into the system from the side where the pink ruby is situated.

Decoupling of the signal and pump channels is ensured by the fact that the planes of polarization of these fields are mutually perpendicular. Then a reduction in the narrow wall of the main waveguide to 4 mm leads to the fact that for the pump this channel becomes a cutoff waveguide. The pump waveguide is, of course, a cutoff waveguide for the signal.

At the pump frequency the structure is a multimode and low-Q-factor resonator. For sufficiently deep saturation of the pumping transition it is necessary to have a power of 150 mW. The overall appearance of the maser is shown in Fig. 19. The input and output waveguides of the signal channel and the pump waveguide are soldered into the sealing flange at the top of the metal cryostat. The waveguide flanges are connected to the case of the slow-wave structure.

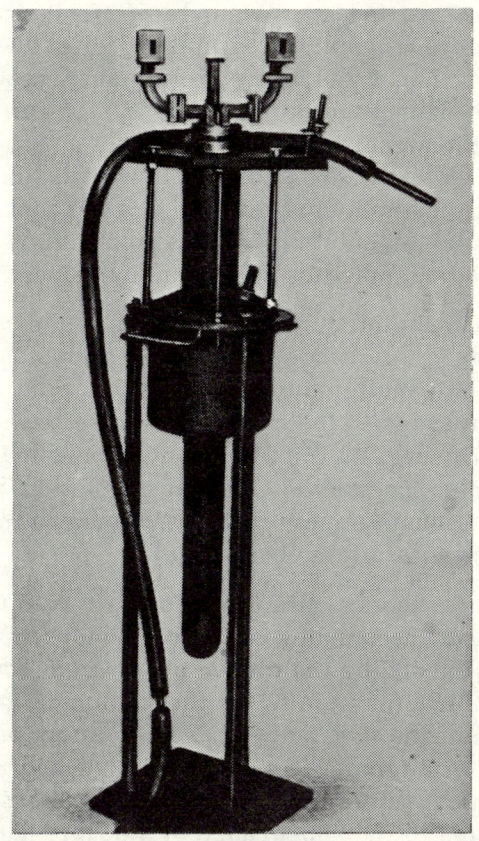

Fig. 19. Over-all appearance of the maser.

We used a metal nitrogen-free cryostat KR-09 [82], which preserved the liquid helium under the operating conditions for 24 hours. The internal diameter of the rear extension of the cryostat is 40 mm and the external diameter 60 mm. To increase the uniformity of the external magnetic field to ± 1 Oe over a length of 100 mm we used shimming rings. The external diameter of the shimming rings was 190 mm, the internal diameter was 172 mm, and the thickness was 5 mm. To measure the uniformity of the field we also used DPPH.

Therefore, the use of post slow-wave structures and uniform ruby rods enabled us to construct a traveling-wave maser at a wavelength of 3.2 cm suitable, in particular, for radio astronomy investigations.

The experience in designing and adjusting this maser leads to the conclusion that the main factor which limits the successful design of a series of masers of this type and in this waveband is the difficulty of obtaining sufficiently uniform crystals of large linear dimensions.

§ 2. Two-Resonator 21-cm Wavelength Maser Intended for Radio Astronomy Investigations

In the previous discussion considerable space was devoted to the theoretical analysis of two-resonator masers which are extremely promising for application in the decimeter band. Their advantages are the relatively large bandwidth for small crystal dimensions, the smaller pumping radiation power than in the case of traveling-wave masers, and high stability.

A two-resonator maser was developed in the laboratory [83, 84] intended for use in the 22-meter radio telescope of the Physical Institute of the Academy of Sciences of the USSR for the observation of the radio line of neutral hydrogen at a wavelength of 21 cm. In this amplifier two strip resonators previously developed by us were combined. The distinctive feature of these resonators is the use of a strip of variable length, which is very convenient for tuning. For this purpose the plane central conductor of the strip resonator (the strip) has a longitudinal hollow inside which there is placed a planar tab. This central tab emerges from the hollow near the open-circuited quarter-wave end of the strip, where the high-frequency current density is small and the quality of the contact does not play any important part. Such a change in tuning of the resonator does not worsen its Q-factor. The strip was soldered with Wood's metal into the waveguide of the pumping radiation with cross section 5 × 17 mm. The wide walls of the waveguide played the part of an external conductor of the strip line. A waveguide resonator at the pumping frequency is formed by the plug which carries the strip and by a short-circuiting piston the movement of which tunes this resonator without affecting the tuning at the signal frequency. The pump resonator is excited by means of a post antenna, which emerges from the pump waveguide feeder. For coupling at the signal frequency a loop antenna is used which is a direct prolongation of the central core of the coaxial signal feeder. When designing the two-resonator maser two such quarter-wave strip resonators were mounted in parallel in a waveguide of square cross section 17 × 17 mm, which serves as a resonator at the pump frequency. To regulate the coupling between the resonators a screened plate is introduced which can be moved for tuning. The input resonator is coupled to the signal feeder by means of a loop, the plane of which coincides with the plane of the resonator strip. We used as the working material ruby with a density of ions of Cr^{3+} of the order of 0.04%, which corresponds to a paramagnetic resonance line width $\Delta\nu_l$ = 50 MHz. In the perpendicular orientation for a magnetic field of 2060 Oe a pumping radiation of wavelengths 2.66 cm corresponds to the 21 cm wavelength signal.

The ruby crystals were cut in the form of Π-shaped plates, closely packed on the strip of the signal resonators and tightly inserted into the square waveguide of the resonator. The over-all dimensions of each crystal were of a rectangular parallelepiped 17 × 17 × 23 mm. In accordance with the field distribution in the resonator and with the matrix elements of the transition operator we chose for the trigonal axis of the crystal the direction perpendicular to the length of the strip and lying in the plane of the strip, and for the external magnetic field the direction parallel to the strip length. In this case one must take precautions to ensure that the axes of the two crystals are parallel to an accuracy to within less than 20'.

The frequency characteristic of the maser, in accordance with the theory given in Chapter IV, was shaped by the symmetrical tuning of the natural frequencies of the resonators in relation to the central frequency of the signal transition line and by the choice of optimum coupling between them. For a detuning of the resonators by 5 MHz ($\Omega = \pm 0.07$) by the choice of optimum coupling for a gain of 18 dB and a sufficiently flat top of the frequency characteristic a passband of 8 MHz was obtained. These results were obtained at a temperature of 4.2°K. An increase in the detuning led to a sharp fall in gain, a reduction in the coupling between the resonators reduced the bandwidth, and an increase in the coupling led to the appearance of a double hump.

The quite sharp fall in the sides of the frequency characteristic is explained by the fact that for the attainable bandwidth the linewidth of the signal transition cannot be assumed to be infinitely large. The theory developed above takes this fact into account.

A comparison of the two-resonator maser with two active resonators and the two-resonator maser of the same construction [85] but with a passive input resonator confirmed that for the same gain the bandwidth of the maser with two active resonators was approximately $\sqrt{2}$ greater than the bandwidth of the maser with a passive input resonator [see formulas (5.66) and (5.63)]. This increase in bandwidth is important for radio-astronomical applications of this maser.

The high stability of the two-resonator maser with two active resonators is also of great importance for radio-astronomical applications. An experimental comparison has shown that, in accordance with the theoretical analysis, with a correctly chosen value of the coupling between the resonators this maser is 3-4 times more stable than a maser with a passive input resonator. For this comparison the intensity of the input noise of the masers was recorded over a long interval of time on an automatic recorder type ÉPP-09. For a gain of 20 dB the permanent instability of the input level was 2-3% for a maser with two active resonators and 8-10% for a maser with a passive input resonator. If the coupling between the resonators does not correspond to the maximally flat characteristic, the instability of the maser with two active resonators does not differ from the instability of the single-resonator maser.

It follows from the previous discussion that the phase and amplitude stability of masers is determined to a considerable extent by the stability of the magnetic field. Therefore, in the maser the field of 2060 Oe required for its operation was produced by a superconducting solenoid closed after the required field is established by a superconducting connector. The choice

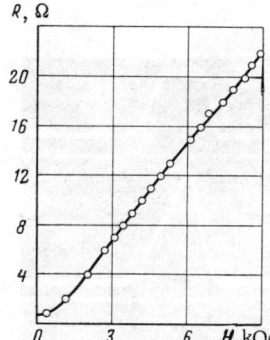

Fig. 20. Calibration of the magnetoresistance.

of the orientation of the crystal axes and the strip resonator enabled us to take the most convenient vertical arrangement of the solenoid from the constructional point of view. The use of a magnet with superconducting windings enabled us to eliminate completely the problem of instabilities due to variations in the magnetic field.

Note that a convenient method of measuring the field of the electromagnet with superconducting windings is the use of the magnetoresistance of a specimen of InSb semiconductor cut in the form of a Corbino disk. We used a disk of thickness 0.5 mm and diameter 5 mm made from n-type InSb with a carrier density of 10^{14} cm^{-3}. The central and peripheral electrodes of the disk are layers of metallic indium. Copper conductors are soldered to the layers of indium. To avoid distortion of the calibration it is necessary to pay special attention to the quality of the soldering.

At room temperature the resistance of the disk was a fraction of an ohm. At liquid helium temperature it was 1 Ω, in a magnetic field of 9 kOe the resistance increased to 22 Ω (Fig. 20). The large slope $\partial R/\partial H$ and the high degree of linearity for $H > 2$ kOe enabled us to measure the magnetic field very accurately, the small thickness enabled us to plot the distribution curve of the field, and the practically complete lack of inertia of the magnetoresistance enabled us to measure not only the average value of the field but the amplitude of its modulation.

To realize the theoretically possible high sensitivity of a radio-astronomical receiver using a maser, one must eliminate parasitic modulation, i.e., the appearance of parasitic signals when the cosmic radiation being investigated is periodically connected to the input of the receiver or the antenna, or when a passive equivalent of the antenna is connected. Such parasitic signals considerably worsen the real sensitivity of the receivers. In a spectral radiometer, i.e., a radiometer intended for investigating the spectral line of cosmic radio radiation, the most dangerous form of parasitic modulation is the periodic deformation of the frequency characteristic of the receiver-amplifier channel. When using a maser in the radiometer such deformation can occur as a result of mismatching the antenna-feeder channel to the frequency characteristic of the maser resonators. The change in the frequency characteristic when switching from the antenna to the antenna equivalent occurs due to the difference which inevitably exists in the matching of these channels.

The theoretical analysis carried out in Chapter IV shows that in fact the mismatch of the antenna-feeder channel leads to a sharp deformation of the frequency characteristic of the maser, the appearance of peaks, dips, etc. This effect is particularly strong if the frequency characteristic of the two-resonator maser is shaped by detuning the resonators relative to one another.

Therefore, before setting up the maser for use in the radio telescope we made additional detailed investigations of this effect under laboratory conditions.

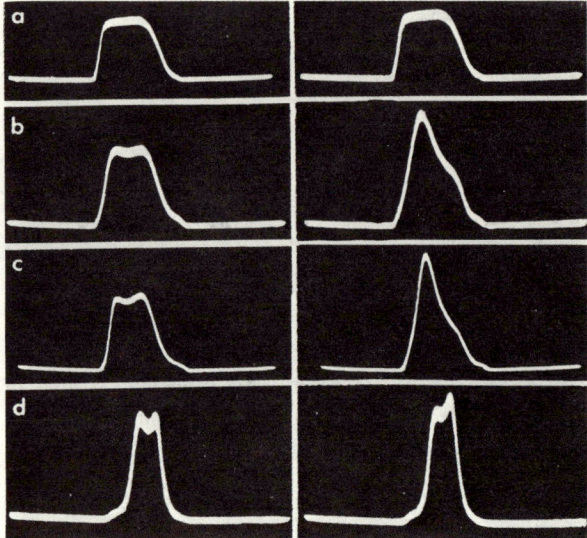

Fig. 21. Experimental frequency characteristics of a maser with two active resonators. a) Gain 10 dB, decoupling 35 dB; b) gain 10 dB, decoupling 20 dB; c) gain 15 dB, decoupling 20 dB; d) gain 20 dB, decoupling 35 dB.

The input of the maser was connected alternately via a circulator and by means of a modulator to loads with different matchings. In these experiments we used loads which had values of VSWR of 1.1 and 1.7 which did not differ very much from one another. The latter of these values exceeds the value of VSWR which is actually obtained when tuning the radio telescope. As follows from formula (4.77) the deformation of the frequency characteristic is essentially determined by the gain and the decoupling. Therefore, the measurements were made for various values of the gain and the decoupling.

Figure 21 shows photographs which illustrate the frequency characteristics obtained experimentally for various values of the gain, decoupling, and mismatch. The curves in the left column correspond to VSWR values of 1.1, and those in the right column correspond to VSWR values of 1.7, other conditions being equal. The results obtained show that the frequency characteristic of two-resonator masers is in fact deformed when the matching of the antenna-feeder channel changes. This deformation is larger the higher the gain and the smaller the decoupling. Comparison of the experimental curves with the calculated curves of Fig. 7 shows that the analysis given in Chapter IV correctly describes the effect of mismatch of the antenna-feeder channel on the operation of a maser and can be used to choose the parameters of the elements of a radio receiver with a maser. When using the above maser in practice in a radio telescope with a gain of 18 dB it is necessary to have a decoupling of 40 dB which requires the use, besides a circulator, of an additional ferrite isolator. As a result, the overall loss in the high-frequency channel was 0.85 dB, of which 0.6 dB is due to losses in the circulator and the isolator.

The inherent noise temperature of the receiver was 35°K, of which 24°K is mixer noise referred to the input of the maser. For losses in the channel of 0.85 dB this led to a contribution to the effective noise temperature of the whole receiver of 42.5°K (see Chapter II). In this case the channel noise temperature was 52.5°K. The noise temperature of the antenna was 40°K. However, it was artificially increased to 85°K in order to bring it to the noise temperature of the antenna equivalent which is at the temperature of liquid nitrogen.

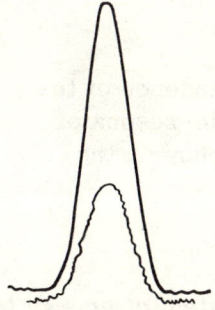
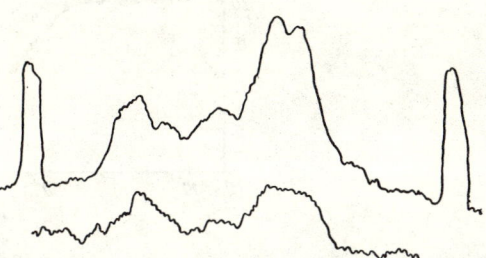

Fig. 22. Two recordings of the radio radiation of a point source in the continuous spectrum. The upper curve was obtained using a maser.

Fig. 23. Two recordings of the radio line of hydrogen. The upper curve was obtained using a maser.

The reduction in the difference in temperatures of the antenna and the antenna equivalent is useful for reducing the technical level of sensitivity (see § 4, Chapter II).

As a result, in accordance with the analysis given in Chapter II, the overall equivalent noise temperature of the receiver as a whole was 180°K. With this noise temperature we realized the theoretically possible sensitivity of the radiometer [53]. In the continuous spectrum with a bandwidth of 8 MHz and a time constant of 1 sec we obtained a fluctuation sensitivity of 0.12°K, and for the observed radio lines of hydrogen a fluctuation sensitivity of 0.35°K (bandwidth 20 kHz, τ = 30 sec). Figures 22 and 23 show examples of recordings in the continuous spectrum and in the radio line of hydrogen. These results show that the use of a maser at a wavelength of 21 cm increased the sensitivity of the radio telescope by a factor of 7-8 in the continuous spectrum and by a factor of 15 in the line of hydrogen. The observations show that the use of a maser enables one to separate out fine detail in the profiles of the radio lines and, consequently, to reveal fine structure in the distribution of interstellar hydrogen. A further improvement in the sensitivity of this radio telescope can be obtained by cooling the circulators and isolators, by reducing their loss, reducing the temperature of the antenna equivalent to 4.2°K, decreasing the antenna noise, and increasing the gain of the maser or reducing the mixer noise.

The whole of this experiment well illustrates the state of the theory of masers and their application, developed in the previous chapters, and clearly shows the prospect of considerably improving the sensitivity of radio receivers by means of them.

§ 3. The Use of Rutile in Masers

Ruby is the most widely used material in masers. Although it is an excellent material for most applications it has a number of drawbacks. The first of these begins to appear when progressing into the region of very short waves. It is very difficult to obtain amplification with ruby at wavelengths shorter than 2-3 cm when using only the Zeeman sublevels of the ground state of chromium ions, since the zero splitting of these sublevels is only 0.383 cm^{-1}. The second drawback of ruby in these applications of masers where it is possible for them to be saturated by the signal field is the quite long spin-lattice relaxation time of this material: $T_1 \approx$ 0.1 sec. As a result, masers using ruby are easily saturated, and the process of restoration of the gain takes a long time. For illustration, Fig. 24 shows an experimentally obtained curve of the gain of a single-resonator maser using ruby in the 10-cm wavelength range. The gain restoration time was 0.5 sec. These results agree well with the results of the theoretical discussion given in Chapter V. In various laboratories experiments have been carried out on new paramagnetic materials, the properties of which enable one to construct masers with better amplitude and relaxation characteristics and with wider bandwidths.

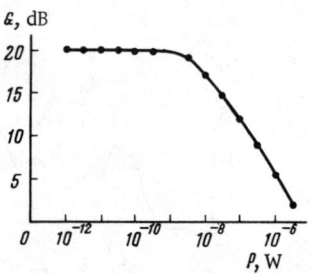

Fig. 24. Dependence of the gain of a single-resonator maser using ruby on the input signal.

Corundum with an impurity of Fe^{3+} ions possesses a system of energy levels [86], which enables us to obtain amplification over a much wider frequency band than corundum with impurity of Cr^{3+} ions (ruby). The development of a maser using this material ($Al_2O_3:Fe^{3+}$) at a wavelength of 3 cm [87], as well as our experiment in the 10-cm range has shown that masers using this material also have a somewhat shorter gain restoration time than an amplifier using ruby. However, this difference is small. In addition, there are serious difficulties in the controlled introduction of the required amount of Fe^{3+} ions into the corundum lattice.

Rutile TiO_2 with Cr^{3+} or Fe^{3+} ion impurity has turned out to be very promising. The large zero splitting (43.3 GHz for chromium, and 43.1 and 83.2 GHz for iron) [88, 89] and the short spin-lattice relaxation time at liquid-helium temperatures and working concentrations (0.5-1 msec) [90] are the reasons for this.

In the millimeter waveband it is easy to obtain inversion in rutile with iron. However, the high dielectric constant of rutile ($\varepsilon_\parallel = 220$ and $\varepsilon_\perp = 120$ at 4.2°K) make it difficult to design masers, since difficult problems arise of matching to the waveguides the many types of oscillations which simultaneously arise in small volumes of this material.

As a rule, in the millimeter waveband dielectric rutile resonators are used. In our experiments on the design of a maser using rutile with iron at a wavelength of 10 mm the resonator was a thin plate of rutile of dimensions $0.5 \times 3 \times 5$ mm placed parallel to the wide wall of the waveguide of cross section 7.2×3.4 mm in the middle of its height. Besides the resonance type oscillations at the required frequencies, coupling to which was controlled to the required degree, good quality oscillations which are poorly coupled to the waveguide are also excited in the specimen at frequencies somewhat remote from the required ones. Therefore, the amplification at the required frequency is accompanied by generation at accompanying frequencies shifted to one side by approximately 30 MHz. This generation can be tuned out in superheterodyne reception, but its presence, of course, reduces the stability of the maser. This effect has been considered by a number of authors [91, 92] who used rutile resonators in the millimeter waveband.

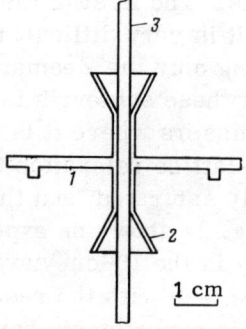

Fig. 25. Sketch of a device for sealing the inlet of the dielectric waveguide in the cryostat.
1) Flange, 2) horn, 3) waveguide.

Note that a convenient experimental method of working in the millimeter waveband is to use flexible dielectric polyethylene waveguides, which considerably simplifies working with sources of pumping radiation in the short-wave part of the millimeter waveband. We also used these waveguides for channelling the pumping energy into the liquid-helium since they withstand liquid-helium temperatures very well. Figure 25 shows a sketch of a double horn which enables us to lead the dielectric waveguide through the sealing flange of the top of the cryostat without loss of microwave energy. Note that the evaporation of the liquid-helium when using dielectric waveguides is less than in the case of metal tubes.

In the decimeter waveband the high dielectric constant which is an obstacle to the effective use of rutile in the millimeter waveband has a number of advantages. These are the reduction in the dimensions of the resonator systems and the increase in the concentration of high-frequency pumping energy per unit volume of the working material. In addition, for rutile when working in the decimeter waveband the ratio of the pumping frequency to the signal frequency is very large, which leads to a high inversion coefficient. Decimeter-wave masers using rutile operate very stably and possess short gain restoration times. Thus, we constructed a single-resonator maser at a wavelength of the order of 10 cm in which we used as the working material rutile with an addition of 0.1% (in the initial mixture) of Cr^{3+} ions [75]. Using the method of pulse saturation we measured that the spin-lattice relaxation time of the EPR line of the signal transition for chromium ions in rutile with a concentration of 0.1% at 4.2°K is $T_1 = 0.5 \cdot 10^{-3}$ sec. When investigating masers with this material we found that after removing the completely saturating action for a nominal gain of 19 dB the gain was restored to the level of 16 dB ($\alpha = 0.7$) in a time $\tau_\alpha = 1.2 \cdot 10^{-3}$ sec, which completely corresponds to formula (5.23).

In this maser the constant magnetic field was produced by an electromagnet with an iron core and superconducting windings made from niobium, which considerably exceeded the stability of its characteristics. Measurements of the dependence of the amplified signal phase on the pumping frequency showed that, in complete accordance with the theory developed above, when the pumping frequency was changed within the limits of the half-width of the line of the pumping transition the phase of the amplified signal to a high degree of accuracy remained unchanged. (This important result was obtained in experiments carried out by Pimenov and Prokhorov, whom the author thanks.)

At wavelengths in the 20-cm band, which are of great interest in radio astronomy, we also carried out experiments which are necessary for the development of masers using chromium-doped rutile as the working material. The results obtained in determining the dependence of the inversion on the EPR linewidth in rutile with chromium on the concentration of Cr^{3+} ions led to extremely encouraging results. The measurements were made with the external magnetic field oriented along the C axis of the crystals. We investigated specimens with chromium ion impurities at concentrations of 0.07, 0.1, 0.15, and 0.3%, which were determined according to the ratio of the number of chromium atoms to the number of titanium atoms. The measurements were made at a temperature of 4.2°K. The magnetic fields which correspond to the signal transition between the levels of the lower doublet in the 20-cm band have values from 1300 to 700 Oe.

The shape of the EPR line of the transition turned out to be intermediate between Lorentzian and Gaussian for all the concentrations investigated, and the linewidth in accordance with the theory of Kittel and Abrahams is proportional to the square root of the concentration. Table 1 shows the results of an experimental determination of the linewidth as a function of the concentration. For comparison we give in the third column the values calculated in accordance with the square root of the concentration proportionality law. We see that for a concentration of 0.3% nonuniform broadening begins to appear.

TABLE 1. Dependence of the EPR Linewidth of the Cr^{3+} Ion on the Concentration

Concentration, %	Measured linewidth, MHz	Calculated linewidth, MHz
0.07	31	31
0.1	37	36
0.15	47	44
0.3	68	62

TABLE 2. Inversion for Different Concentrations

Concentration, %	Inversion	Product of the inversion and concentration
0.07	20	0.14
0.1	20	0.2
0.15	12	0.18
0.3	−1	−0.03

The inversion was determined for a pumping radiation frequency of 47 Hz. The results obtained are shown in Table 2. Since the value of the Q-factor of the active material is determined by the product of the inversion and the concentration, we give this value in the third column of Table 2.

The data presented show that the optimum density is 0.1%. A knowledge of the inversion, the concentration, and the line width enables us to calculate the Q-factor of the active material. For the optimum concentration and complete filling an estimate of the Q-factor of the active material gives a value of $|Q_B| = 80$. As a check we constructed a simple single-resonator reflection maser, the resonator of which is a silver-plated parallelepiped cut from rutile with an impurity of 0.1% of chromium ions. Measurement of the bandwidth and gain give a value of $|Q_B| = 70$, which confirms the correctness of the results obtained from the measurement of the inversion. We note here that the high dielectric constant of rutile enabled us to use a volume resonator with complete filling in the decimeter waveband. The data presented show the undoubted promise of the use of rutile with chromium as the active material of paramagnetic masers in the 20-cm band.

To conclude this section we mention that the difficulties in the use of rutile for masers in the millimeter waveband have stimulated the search for new paramagnetic materials suitable for use in this band. We have investigated from this point of view the EPR spectra of Cr^{3+} ions in monoclinic crystals of $ZnWO_4$ and $CdWO_4$ and we have determined the spin Hamiltonian constants for these materials. Table 3 shows values of the zero splittings and indicates the orders of magnitude of the spin-lattice relaxation times T_1 (measured by Manenkov and Milyaev) at 4.2°K for the chromium ion in the rutile lattice of magnesium tungstate $MgWO_4$ [93] and zinc and cadmium tungstate $ZnWO_4$ and $CdWO_4$. The dielectric constant, according to our data, for $ZnWO_4$ has an order of magnitude of 8-10, and for $CdWO_4$ is 15-20, which, of course, is considerably less than for rutile. Comparison of the properties of these materials clearly indicates the advantages of using tungstates in millimeter-band masers.

TABLE 3. Zero Splittings and Relaxation Times for the Chromium Ion in Rutile and Tungstates

Crystal	Zero splitting, GHz	$4.2°K \cdot 10^{-3}$ sec	Concentration, %
TiO_2	43.3	0.5	0.1
$MgWO_4$	48.1	0.8	0.1
$ZnWO_4$	51.5	1.1	0.1
$CdWO_4$	86.1	0.4	0.03

We mention also that a specially conducted investigation of the optical pumping of F-centers in KBr showed that for pumping by nonmonochromatic light the efficiency of obtaining inversion of the Zeeman sublevels of the ground state is small, although inversion was obtained in fields from 700 to 4000 Oe when the specimen is illuminated by light at a wavelength of 17,450 cm^{-1} with a spectral band of 200 cm^{-1} and an intensity of several microwatts.

§4. Investigation of Calcium-Vanadium Ferrite as a Material for Low-Temperature Nonreciprocal Elements of Masers

The analysis carried out in the first chapters of this work and tests on the operation of a two-resonator maser at a wavelength of 21 cm on the large radio telescope of the Physics Institute of the Academy of Sciences of the USSR have enabled us to draw conclusions on the need to design masers provided with nonreciprocal elements cooled to liquid-helium temperatures. In this case for resonator masers both circulators and isolators are necessary. To construct these, ferrites are required which preserve their activity when cooled to liquid-helium temperatures. However, when the majority of materials are cooled the ferromagnetic resonance line is broadened up to several thousand Oersted, the saturation magnetization increases and the activity is therefore lost. As a rule, satisfactory results can be obtained when using yttrium-iron garnets with gadolinium and without it. Nickel-zinc ferrite is also used [94]. However, these materials are technically very difficult to purify from parasitic impurities of the rare-earth ions and bivalent iron which lead to a broadening of the line. In turn, line broadening is the cause of an increase in the forward losses. For nonreciprocal elements of traveling-wave masers this increase in the forward losses is not as important as for the circulators of resonator masers.

Any considerable reduction in the forward losses in cooled circulators using yttrium garnet requires surmounting great technological difficulties. It is therefore of interest to synthesize calcium-vanadium ferrite with a garnet structure containing neither yttrium nor iron [95, 96]. Its chief merit is the ease of technological purification and manufacture. The Curie temperature is 493°K, and we might therefore expect that when it is cooled from room temperature to liquid-helium temperature its properties will not change very much. In view of the absence of any data on the properties of CaV-ferrite at liquid-helium temperatures, we carried out an experimental investigation of the ferromagnetic resonance of the ferrite. We investigated polycrystalline specimens manufactured under technologically controlled conditions, cut in the form of spheres of 1.5 mm diameter. The specimens were placed in a strip resonator at a wavelength of 18 cm.

The dependence of the ferromagnetic resonance linewidth and the resonance field on the temperature is shown in Fig. 26. The comparatively small change in the values of the linewidth and the resonance field emphasizes the advantage of using this material. Essentially, we can construct a circulator with CaV-ferrite at room temperature and operate without readjustment at liquid-nitrogen and liquid-helium temperatures.

Based on these results, we used CaV-ferrite to design a circulator at a wavelength of 18 cm. A disk cut from the ferrite was placed between crossed strip lines. We used a standard

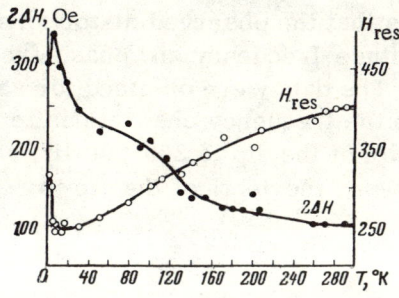

Fig. 26. Dependence of the linewidth and resonance field of calcium-vanadium ferrite on temperature.

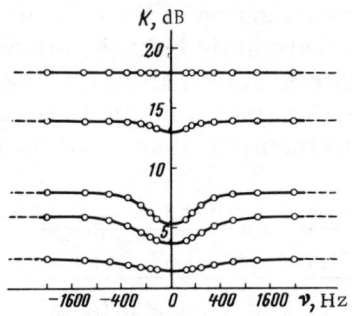

Fig. 27. Experimentally measured distortion of the amplitude–frequency characteristic.

coaxial excitation. The over-all dimensions of this circulator corresponded to the standard glass helium cryostat. The parameters obtained as a result of the design confirmed the conclusion on the advantages of this material. At 300°K we obtained a decoupling of 35 dB for forward losses of 0.8 dB, and at 4.2°K decoupling of 30 dB for forward losses of 1.2 dB. The bandwidth at the 20 dB level was 60-100 MHz.

Calcium-vanadium ferrite was also used to design a Y-circulator intended for operation in conjunction with a 3-centimeter maser [76]. At 4.2°K we obtained decoupling of 46 dB for forward losses of 0.8 dB and a bandwidth of 170 MHz. When operating in conjunction with the maser at liquid-helium temperature the circulator increased the noise temperature of the maser by only 0.7°K, which is undoubtedly an important achievement.

Therefore, calcium-vanadium ferrite with a garnet structure is in fact a suitable material for constructing circulators which are cooled to liquid-helium temperatures intended for operation in conjunction with masers.

§5. Experimental Investigation of Nonlinear Distortions in a Paramagnetic Maser

As already stated earlier, when using masers in communication lines the linearity of the maser is very important. Therefore, the theoretical account given in Chapter V for a traveling-wave maser was supplemented by an experimental investigation of a 3-cm resonator maser. We used rutile with iron ions as impurity as the working material. The nominal gain was 18 dB (G_0 = 63) for a bandwidth at the 3 dB level of 6 MHz. The operating temperature was 4.2°K. Stability of operation of the maser was ensured by using a high-power pumping source and a magnet with superconducting windings. A signal at the central frequency of the passband was fed to the input of the maser. Modulation (5%) of this signal was carried out by an external diode modulator to which rectangular pulses of a control voltage were fed. The amplified signal from the output of the maser was fed to the input of a traveling-wave amplifier at the output of which there was a video detector. The output signal was recorded on an oscilloscope type ÉNO-1. When the intensity was increased the shape of the output pulses was distorted. The linearity of the traveling-wave amplifier and the square-law characteristic of the video detector were specially checked.

Spectral analysis of the output pulses shows that the observed distortions of the shape of the pulses correspond to the changes in the amplitude-frequency and phase-frequency characteristics of the maser shown in Figs. 27 and 28. The data were obtained for saturation for which the gain is 14, 8, 6, and 2.5 dB. The dip in the frequency characteristic reaches a maximum value of 2.5 dB, for a gain of 8 dB. The width of the dip is 200-300 Hz, and the maximum phase shift is ±22°. When the saturation is increased the depth of the dip is reduced but the width increases.

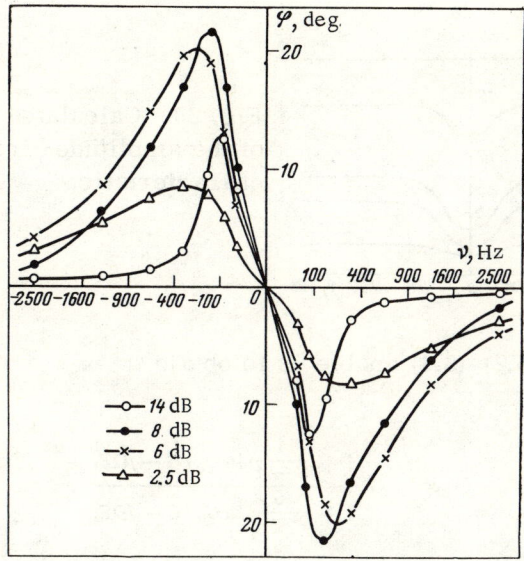

Fig. 28. Experimentally measured distortion of the phase−frequency characteristic.

The spin-lattice relaxation time T_1 for the material used is $3 \cdot 10^{-3}$ sec. The presence of a dip and the form of the observed behavior correspond to the results of the theory and confirm the conclusion that practically over the whole bandwidth of paramagnetic masers the nonlinear distortions are small. It is useful to carry out a more careful comparison of the results of the experimental investigation and the theoretical analysis.

For square-law detection we recorded on the screen of the oscilloscope the envelope of the field intensity at the maser output. We will introduce the effective gain of the envelope

$$G_{\text{eff}} = \frac{\delta P_{\text{out}}}{\delta P_{\text{in}}}, \qquad (6.1)$$

where δP_{in} and δP_{out} are the assumed modulation changes of the square of the amplitude of the signal at the input and output of the maser respectively.

For small modulation

$$G_{\text{eff}} = \frac{V_{\text{out}}}{V_{\text{in}}} \frac{V_{1\text{out}}}{V_{1\text{in}}}, \qquad (6.2)$$

where V_{out} and V_{in} are the amplitudes of the strong field at the resonant frequency, and $V_{1\text{out}}$ and $V_{1\text{in}}$ are the amplitudes of the sidebands at the output and input respectively.

For amplitude modulation with frequency modulation $\Omega \ll 1/T_2$ the expression (5.61) for $V_{1\text{out}}$ can be represented in the form

$$\frac{V_{1\text{out}}}{V_{1\text{in}}} = G^{1/2} \frac{1 + SP - j\Omega T_1}{1 + GSP - j\Omega T_1}. \qquad (6.3)$$

At the same time we can write

$$\frac{V_{\text{out}}}{V_{\text{in}}} = G^{1/2}, \qquad (6.4)$$

where the expression for G is given by formula (5.13).

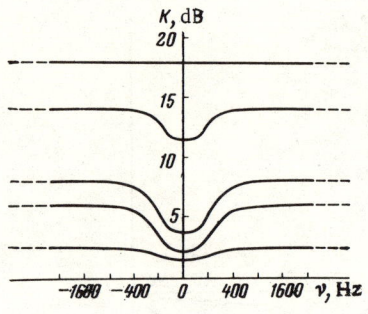

Fig. 29. Calculated distortion of the amplitude–frequency characteristic.

Relations (5.13) and (6.2)-(6.4) enable us to obtain

$$G_{\text{eff}} = G \frac{1 + \frac{1}{G-1} \ln G_0/G - j\Omega T_1}{1 + \frac{G}{G-1} \ln G_0/G - j\Omega T_1}. \tag{6.5}$$

The following amplitude and phase characteristics correspond to formula (6.5)

$$|G_{\text{eff}}| = G \sqrt{\frac{\left(1 + \frac{1}{G-1} \ln G_0/G\right)^2 + \Omega^2 T_1^2}{\left(1 + \frac{G}{G-1} \ln G_0/G\right)^2 + \Omega^2 T_1^2}} \tag{6.6}$$

and

$$\tan \varphi = \frac{-\Omega T_1 \ln G_0/G}{\left(1 + \frac{1}{G-1} \ln G_0/G\right)\left(1 + \frac{G}{G-1} \ln G_0/G\right) + \Omega^2 T_1^2}, \tag{6.7}$$

which are shown in Figs. 29 and 30. For ease of comparison with the experimental data the curves of Figs. 29 and 30 are calculated for G_0 = 18 dB for G = 14, 8, and 2.5 dB. The time T_1 was taken equal to $3 \cdot 10^{-3}$ sec.

A comparison of the calculated data with the experimental data shows that the theory developed in Chapter V enables us to make a fairly reliable estimate of the nonlinear distortions in masers. We recall that the theory was developed for a traveling-wave maser and the experiments were carried out for a resonator maser.

Conclusion

In this work we have presented the results of a series of experimental and theoretical investigations on masers, in the course of which we have obtained the limiting sensitivity of receivers of electromagnetic radiation, clarified the gain in sensitivity when masers are incorporated in receiving systems, and determined the important properties of masers, such as the bandwidth, stability, the dependence of the gain on the signal power, the action of the gain restoration time after the saturating signal is removed, and the value of the nonlinear distortion.

Masers successfully combine high sensitivity with stability, and linearity. Their use has enabled us for the first time in the history of microwave electronics to reduce the internal noise of receivers to such extremely low levels that the sensitivity is determined by the external noise.

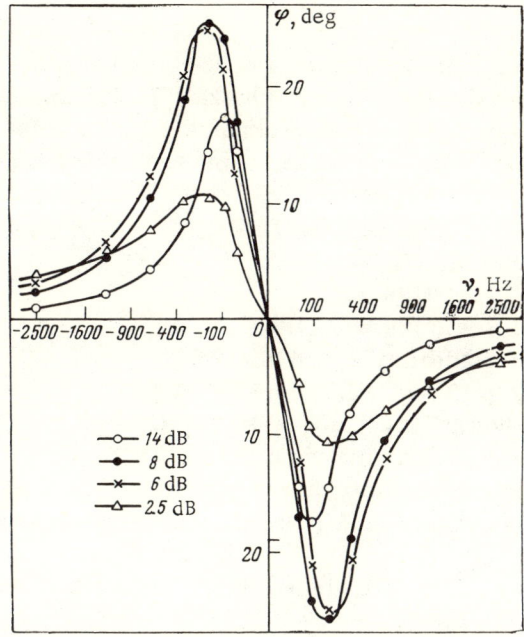

Fig. 30. Calculated distortion of the phase-frequency characteristic.

Masers are already being used in radio astronomy, radar, and global and space communication links. It is useful, however, to point out certain problems which require further investigation in connection with the application of masers. First of all, for certain applications it is desirable to develop masers with a large bandwidth. Then, in the centimeter and decimeter wavebands it is necessary to use traveling-wave masers, masers with coupled resonators, and multiresonator systems of the band-pass filter type. In certain cases masers are required which possess a large dynamic range and small gain restoration time after saturation by a strong signal. It is then necessary to use in the masers working materials with fairly short relaxation times. One of these is rutile with ions of chromium or iron.

The use of masers requires the maximum reduction in the noise of the antenna-feeder channels. Their use then provides an enormous advantage. To reduce the channel noise one should use circulators, isolators, switches, etc., cooled to liquid-helium temperatures.

To realize the high stability of masers they should be used with magnets with superconducting windings. To improve the operating and technical characteristics it is extremely important to use closed-cycle helium liquifiers.

The possibility of developing masers in the millimeter and submillimeter band is of great interest.

The author thanks A. M. Prokhorov, whose stimulating influence to a considerable extent determined the content of this work. The author also thanks I. A. Abramenkov, G. N. Artemychev, E. K. Karlova, B. B. Krynetsko, T. I. Kuznetsova, I. A. Skuratov, and A. D. Shirkov, whose investigations contributed to this work.

References

1. N. V. Karlov and A. M. Prokhorov, Radiotekhnika i Élektronika, 9:2088 (1964).
2. N. V. Karlov and A. M. Prokhorov, Radiotekhnika i Élektronika, 7:328 (1962).
3. N. V. Karlov and B. M. Chikhachev, Radiotekhnika i Élektronika, 4:1048 (1949).
4. C. H. Townes and R. Serber, Quantum Electronics, New York (1960), p. 233.
5. H. Heffner, Proc. IRE, 50:1604 (1962).
6. G. S. Gorelik, Uspekhi Fiz. Nauk, 44:348 (1951).
7. J. Weber, Phys. Rev., 108:537 (1957).
8. R. V. Pound, Ann. Phys., 1:24 (1957).
9. M. W. Strandberg, Phys. Rev., 106:617 (1957).
10. F. V. Bunkin, Izv. Vuzov. Radiofizika, 4:496 (1961).
11. A. M. Prokhorov, Zh. Éksp. Teor. Fiz., 40:1384 (1961).
12. L. Mandel, Proc. Phys. Soc., 712(Pt. 6)(468):1037 (1958).
13. L. Hyvarinen, Acta Polytech. Scandinavica, 2:28 (1958).
14. N. G. Basov and N. V. Karlov, Radiotekhnika i Élektronika, 5:676 (1960).
15. N. V. Karlov, Yu. P. Pimenov, and A. M. Prokhorov, Radiotekhnika i Élektronika, 5:676 (1960).
16. G. M. Zverev, N. V. Karlov, L. S. Kornienko, A. A. Manenkov, and A. M. Prokhorov, Uspekhi Fiz. Nauk, 77:61 (1962).
17. S. M. Rytov, Theory of Electrical Fluctuations and Thermal Radiation, Izd. AN SSSR, Moscow (1953).
18. Centimeter Wave Antennas, Sov. Radio, Moscow (1950).
19. E. H. Putley, Proc. IEEE, 51:1412 (1963).
20. V. V. Kobelev, Thesis, FIAN, Physics Faculty, Moscow State University (1952).
21. D. G. Lampard, Proc. IEEE, Pt. 4, 101(6): 118 (1954).
22. F. V. Bunkin, Doctoral Dissertation, FIAN (1962).
23. F. V. Bunkin and N. V. Karlov, Zh. Tekh. Fiz., 25:733 (1955).
24. N. V. Karlov, Candidate's Dissertation, FIAN (1956).
25. K. E. Machin, M. Ryle, and D. D. Vonberg, Proc. IEEE, Pt. III, 99: 177 (1952).
26. N. V. Karlov, Proceedings of the 5th Conference on Problems of Cosmogony, Izd. AN SSSR, Moscow (1956), p. 88.
27. N. V. Karlov and A. E. Salomonovich, Pribory i Tekh. Éksper., 1956(2):105 (1956).
28. N. V. Karlov, Radiotekhnika i Élektronika, 3:74 (1958).
29. J. V. Jelley and B. F. C. Cooper, Rev. Sci. Instr., 32:166 (1961).
30. V. S. Troitskii, Zh. Tekh. Fiz., 25:1426 (1955).
31. N. V. Karlov, Radiotekhnika i Élektronika, 1:852 (1956).
32. N. V. Karlov, Radiotekhnika i Élektronika, 11:271 (1966).
33. A. Veilsteke, Fundamental Theory of Masers [Russian translation], IL, Moscow (1963).
34. V. M. Kontorovich and A. M. Prokhorov, Zh. Éksp. Teor. Fiz., 33(6):1428 (1957).
35. P. W. Anderson, J. App. Phys., 28:1049 (1957).
36. P. Butcher, in: Paramagnetic Masers [Russian translation], IL, Moscow (1961), p. 106.
37. A. Clogston, in: Paramagnetic Masers [Russian translation], IL, Moscow (1961), p. 70.
38. N. V. Karlov and A. A. Manenkov, Izv. Vuzov. Radiofizika, 7(1):5 (1964).
39. N. V. Karlov and A. M. Prokhorov, Radiotekhnika i Élektronika, 8(3):453 (1963).
40. N. V. Karlov and A. M. Prokhorov, Radiotekhnika i Élektronika, 11:267 (1966).
41. N. V. Karlov, R. M. Martirosyan, and R. L. Sorochenko, Radiotekhnika i Élektronika, 10(1):40 (1963).
42. N. V. Karlov and R. M. Martirosyan, Radiotekhnika i Élektronika, 10(4):673 (1965).
43. P. N. Butcher, Proc. IEE, 103B:301 (1956).
44. A. M. Prokhorov, Zh. Éksp. Teor. Fiz., 34:1658 (1958).

45. L. A. Kulevsky, P. P. Pashinin, and A. M. Prokhorov, Proc. 3rd International Conference on Quantum Electronics, p. 1064.
46. J. E. Geusic and H. E. D. Scovil, Bell Syst. Tech. J., 41:1371 (1962).
47. N. G. Basov, A. Z. Grasyuk, and I. G. Zubarev, Dokl. Akad. Nauk SSSR, 157:1084 (1964).
48. V. B. Shteinshleiger, Radiotekhnika i Élektronika, 4:1947 (1959).
49. V. B. Shteinshleiger, G. S. Misezhnikov, and D. A. Afanas'ev, Radiotekhnika i Élektronika, 7:874 (1962).
50. A. W. Nagy and G. C. Friedman, Proc. IRE, 50:2505 (1962).
51. V. B. Shteinshleiger and G. S. Misezhnikov, Radiotekhnika i Élektronika, 9:2099 (1964).
52. Tube Amplifiers [Russian translation], Sov. Radio, Moscow (1950).
53. R. M. Martirosyan, R. L. Sorochenko, and A. M. Prokhorov, Dokl. Akad. Nauk SSSR, 156(6):1326 (1964).
54. J. J. Cook, L. G. Gross, M. E. Bair, and R. W. Terhune, Proc. IRE, 49:768 (1961).
55. N. V. Karlov, Yu. P. Pimenov, and A. M. Prokhorov, Radiotekhnika i Élektronika, 6:411 (1961).
56. N. V. Karlov and T. I. Kuznetsova, Radiotekhnika i Élektronika, 12(2):284 (1967).
57. L. L. Moskvitin and Yu. E. Naumov, Radiotekhnika i Élektronika, 10(12):2105 (1964).
58. E. O. Schulz-Du Bois, Proc. IEEE, 52:644 (1964).
59. W. J. Tabor, F. S. Chen, and E. O. Schulz-Du Bois, Proc. IEEE, 52:656 (1964).
60. F. Bosch, H. Rothe, and E. O. Schulz-Du Bois, Proc. IEEE, 52:1244 (1964).
61. F. Bosch, Z. Angew. Phys., 18:254 (1965).
62. T. I. Kuznetsova and S. G. Rautian, Zh. Éksp. Teor. Fiz., 49(11):1605 (1965).
63. D. N. Klyshko, Yu. S. Konstantinov, and V. S. Tumanov, Izv. Vuzov. Radiofizika, 8:517 (1965).
64. E. K. Karlova, N. V. Karlov, A. M. Prokhorov, and E. G. Solov'ev, Pribory i Tekh. Éksper., 1963(2):107 (1963).
65. A. I. Agranovskaya, N. V. Karlov, and B. B. Krynetskii, Fiz. Tverd. Tela, 8:1265 (1966).
66. N. V. Karlov, Yu. P. Pimenov, and A. M. Prokhorov, Radiotekhnika i Élektronika, 6:846 (1961).
67. N. V. Karlov and B. B. Krynetskii, Fiz. Tverd. Tela, 7:1865 (1963).
68. E. N. Emel'yanova, N. V. Karlov, A. A. Manenkov, V. A. Milyaev, A. M. Prokhorov, S. P. Smirnov, and A. V. Shirkov, Zh. Éksp. Teor. Fiz., 44:868 (1963).
69. E. V. Andreeva, N. V. Karlov, A. A. Manenkov, V. A. Milyaev, and A. V. Shirkov, Fiz. Tverd. Tela, 6:1649 (1964).
70. N. V. Karlov, J. Margerie, and V. Merle D'Aubigne, J. de Phys., 24:717 (1963).
71. N. V. Karlov, T. I. Kuznetsova, B. B. Krynetskii, and A. V. Shirkov, Radiotekhnika i Élektronika, 12(2):364 (1967).
72. V. A. Danilychev, N. V. Karlov, B. D. Osipov, A.V. Shirkov, G. I. Shlippe, Pribory i Tekh. Éksper., 1963(5):221 (1963).
73. N. V. Karlov, Radiotekhnika i Élektronika, 6:1028 (1961).
74. E. M. Dianov, N. A. Irisova, and N. V. Karlov, Pribory i Tekh. Éksper., 1965(4):144 (1965).
75. Yu. P. Pimenov and A. M. Prokhorov, Radiotekhnika i Élektronika, 8(9):1643 (1963).
76. B. B. Krynetskii, G. P. Kuz'min, and A. V. Shirkov, Radiotekhnika i Élektronika, 11(12):2248 (1966).
77. G. Kikuchi, J. Lamb, and R. W. Terhune, J. App. Phys., 30:1061 (1959).
78. R. W. DeGrasse and H. E. D. Scovil, Bell Syst. Tech. J., 38:2 (1959).
79. E. G. Solov'ev and E. K. Karlova, Radiotekhnika i Élektronika, 6(3):406 (1961).
80. V. B. Shteinshleiger and G. S. Misezhnikov, Pribory i Tekh. Éksper., 1959(6):133 (1959).
81. G. Haddad and J. Rowe, IRE Trans. MTT-10:3 (1962), MTT-12:406 (1964).
82. A. B. Fradkov, Pribory i Tekh. Éksper., 1961(4):170 (1961).

83. R. M. Martirosyan and A. M. Prokhorov, Radiotekhnika i Élektronika, 9(12):2099 (1964).
84. R. M. Martirosyan, Candidate's Dissertation, Physics Institute of the Academy of Sciences of the USSR (1964).
85. R. M. Martirosyan and A. M. Prokhorov, Pribory i Tekh. Éksper., 1964(1):106 (1964).
86. L. S. Kornienko and A. M. Prokhorov, Zh. Éksp. Teor. Fiz., 40:805 (1957).
87. L. S. Kornienko and A. M. Prokhorov, Zh. Éksp. Teor. Fiz., 36:919 (1959).
88. H. J. Gerritsen, S. E. Harrison, and H. Lewis, J. App. Phys., 31:1566 (1960).
89. D. L. Carter and A. Okaya, Phys. Rev., 188:1485 (1960).
90. A. M. Manenkov, V. A. Milyaev, and A. M. Prokhorov, Fiz. Tverd. Tela, 4:833 (1962).
91. S. Foner and L. R. Momo, J. Appl. Phys., 31:742 (1960).
92. D. L. Carter, J. Appl. Phys., 32:2541 (1961).
93. V. A. Atsarkin, É. A. Gerasimova, I. G. Matveeva, and A. V. Frantsesson, Zh. Éksp. Teor. Fiz., 43(4):1273 (1962).
94. V. B. Shteinshleiger, Radiotekhnika i Élektronika, 7:1253 (1962).
95. G. A. Smolenskii, V. P. Polyakov, and V. M. Yudin, Izv. Akad. Nauk SSSR, ser. Fiz., 25(11):1396 (1961).
96. S. Geller, G. P. Espinosa, W. J. Williams, R. C. Sherwood, and E. A. Nesbitt, Appl. Phys. Lett., 3:60 (1963).

EXPERIMENTS ON ELECTRON PARAMAGNETIC RESONANCE AT TEMPERATURES OF 0.1-4.2°K

V. B. Federov

Introduction

Electron paramagnetic (spin) resonance, discovered by E. K. Zavoiskii in 1944, is used widely in solid-state physics. The application of EPR spectroscopy has been highly successful in obtaining information on the paramagnetism of crystals. Detailed studies have been carried out on the absorption spectra of ions of the iron and rare-earth groups which had been introduced as isomorphous impurities into the lattices of various diamagnetic crystals. These studies, combined with investigations of the optical absorption in paramagnetic crystals, are the basis of the crystal field theory. The electron paramagnetic resonance (EPR) method has also played a decisive role in investigations of the paramagnetic relaxation in solids. Studies of the mechanisms of relaxation processes, associated with internal interactions in crystals, have yielded detailed information on the spin-spin and spin-lattice interactions of paramagnetic ions in crystals. A practical result of the studies of the paramagnetism of ionic crystals is the development of low-noise quantum paramagnetic amplifiers of weak radio-frequency signals; these amplifiers were the first devices in which the stimulated emission (maser effect) and absorption of radiation in solids were employed for practical purposes. In its turn, the development of these amplifiers has catalyzed numerous studies of the internal interactions in paramagnetic crystals at low temperatures. The use of liquid helium for cooling has made it possible to study spectra and relaxation processes in matter at temperatures down to 1.3°K.

Recently, new and interesting problems have arisen and their solution has required the extension of the EPR experiments in the direction of extremely low temperatures (T < 1°K). Two of these problems are the experimental investigation of the magnetic properties of molecular crystals of chemically stable organic free radicals, and that of the influence of exchange interactions on the paramagnetic spin-lattice relaxation in crystals. These two problems are the subject of the present paper.

Measurements of the static magnetic susceptibility at low temperatures (20-1,5°K) of some organic free radicals (in particular, of α,α-diphenyl-β-picrylhydrazyl, which was investigated in the study reported here) indicated the presence of a weak antiferromagnetic exchange interaction. The corresponding Curie constant Θ (also known as the Weiss constant) is about -1°K. Since the spin systems of such radicals are characterized by a high (close to $f = 1$) concentration of magnetic particles and, consequently, by a regular space structure, it is found that at sufficiently low temperatures (T < 1°K) the exchange interaction affects strongly the energy spectrum of the spin system and may give rise to a transition to an ordered magnetic state. Such a transition should be accompanied by a modification of the system of energy levels in a crystal and, therefore, it can be deduced from changes in the EPR absorption spectrum. The high sensitivity of the EPR method makes it possible to carry out experiments on small samples. The existence of a phase transition from the paramagnetic to the antiferromagnetic state in molecular crystals of free radicals is a very likely event according to the current knowledge of the nature of magnetic properties, and this phenomenon is not trivial because organic free radicals are a relatively new class of substances whose magnetic properties are not yet very well known.

When we investigate the magnetic properties of organic free radicals we must bear in mind a characteristic property of the spin systems of these radicals which distinguishes them from, for example, ionic crystals. This property is associated with a strong delocalization of unpaired electrons in the radical molecule. Consequently, the g factor of the EPR line differs very little from the value g = 2.0023, representing free spins. For example, at 4.2°K, the difference $(g-2)$ is 0.0052 for α,α-diphenyl-β-picrylhydrazyl (DPPH) [1]. The low value of $(g-2)$ of free radicals indicates that the spin-orbit interaction of unpaired electrons is weak and, therefore, the anisotropic properties of the spin system (the g-factor anisotropy and the anisotropic exchange interaction) are also weak. In fact, according to [1], the difference $(g-2)$ changes by only ± 0.0015 when the orientation of a DPPH crystal is varied relative to an external magnetic field. These simple properties of the spin systems of radicals make them attractive subjects for the experimental verification of the various theoretical models of spin systems, which are usually based on the free-spin representation.

There have been only a few experimental investigations of the antiferromagnetic exchange interaction in organic free radicals. The present paper reviews these investigations, describes new experiments in which a transition of the DPPH radical from the paramagnetic to the antiferromagnetic state was observed for the first time, and compares the method used in these experiments, as well as the results obtained, with the methods and results reported by others.

The problem of the influence of exchange interactions on the paramagnetic spin-lattice relaxation in crystals is of interest because of low-temperature anomalies in the one-phonon paramagnetic spin-lattice relaxation.

The theory of paramagnetic relaxation due to de Krönig [2] and Van Vleck [3] deals with the one-particle spin-lattice relaxation. Therefore, the probability of a one-phonon nonradiative transition is independent of the concentration of magnetic particles in a system; the probability of a relaxation transition depends linearly on the temperature even when the ground state of a magnetic particle is a Kramers doublet and the probability is proportional to ν^4, where ν is the resonance absorption frequency. The one-particle Van Vleck–de Krönig model has been confirmed experimentally in studies of magnetically dilute systems at high frequencies, when the Zeeman energy is much higher than the energy of spin-spin interactions.

However, many experiments on magnetically concentrated ($f \approx 1$) and fairly dilute ($f \approx 10^{-2} - 10^{-3}$) systems have yielded results contradicting the Van Vleck–de Krönig theory. The most important contradictions are to be found in the concentration dependence of the spin-lattice relaxation rate and the dependences of the relaxation time on the frequency and temperatue, which are different from those predicted by the theory.

The concentration dependence indicates that we must include spin-spin interactions in the relaxation mechanism. The first step in this direction has been made for magnetically concentrated spin systems with $f \approx 1$, such as the free radical DPPH. The Van Vleck–de Krönig mechanism is ineffective in DPPH because of a weak spin-orbit coupling. Bloembergen and Wang [4] have suggested that the excitation energy is transferred from the Zeeman levels to the spin-

spin degrees of freedom and then to the lattice. This mechanism requires a spin-spin energy comparable with or even higher than the Zeeman energy. In DPPH, this energy is due to the isotropic exchange interaction (the magnetic dipole interaction of spins in the EPR experiments is usually weaker than the Zeeman interaction). This mechanism accounts for the experimentally observed temperature and concentration dependences of the relaxation time in spin systems with high magnetic particle concentrations ($f \approx 1$).

The present paper describes measurements of the temperature dependence of the relaxation time of DPPH in the range T = 4.2-1.5°K. The experimental data obtained are compared with the current theory of relaxation processes.

However, in magnetically dilute systems, the spin-spin exchange interactions are included in the spin-lattice relaxation in a different way. This is because the spin systems of magnetically dilute crystals do not have a regular space structure but represent random distributions of small numbers of magnetic particles over the lattice sites. The distances between the majority of the particles amount to several lattice constants and electron exchange between such particles is impossible. The exchange interaction is appreciable only for a small fraction of the total number of paramagnetic ions, i.e., for those ions which are located at neighboring or nearby lattice sites. If the exchange interaction of a pair of such closely spaced ions is stronger than the interaction of that pair with an external magnetic field, the pair has a spectrum which is different from that of the majority of ions not coupled to each other by the exchange interaction. Van Vleck [5] was the first to draw attention to the fact that ion pairs can interact strongly with the thermal motion of the lattice. In view of this interaction, even widely spaced ions can play an important role in the spin-lattice relaxation.

There have been several investigations of the influence of ion pairs on the spin-lattice relaxation of isolated ions in magnetically dilute systems. For example, two possible types of nonradiative transition involving ion pairs are discussed theoretically in [6-8]. Gill [9] investigated experimentally the relaxation mechanism involving ion pairs. Other experimenters simply reported anomalous (with respect to the Van Vleck−de Krönig model) dependences of the rate of relaxation on the concentration, frequency, and temperature.

Gill measured the rate of relaxation of ion pairs in ruby with a high concentration of chromium ions, as well as in ruby with a low concentration of such ions. He demonstrated experimentally that: a) a thermal equilibrium is established rapidly between pairs and single ions if the concentration of chromium is sufficiently high ($f > 0.21\%$); b) the absolute value of the relaxation rate of pairs is sufficiently high to account for the observed relaxation rate of single chromium ions. It follows from Gill's experiments that, at sufficiently high chromium concentrations ($f > 0.21\%$), ion pairs participate in the transfer of energy from single chromium ions to the lattice; alternately, we can say that such a transfer of energy does not contradict the experimental data.

To demonstrate conclusively the occurrence of this spin-lattice relaxation mechanism, we must know the energy spectrum of ion pairs, find those quantum transitions in the pair spectrum which are responsible for the observed relaxation, calculate theoretically the probability of such transitions, and compare these calculations with the experimental data. This is difficult to do in the case of ruby because of the great complexity of the spectrum of ion pairs [10]. Therefore, we must use simpler systems. A simple and suitable system of this kind is iron-doped potassium cobalticyanide, $K_3(Co,Fe)(CN)_6$.

The purpose of our experiments on this ferrocyanide was to carry out a detailed analysis of the mechanism and to compare quantitatively the theory with the experiment in order to demonstrate the decisive role of ion pairs in the spin-phonon interaction of single ions. The experiments were carried out at relatively low frequencies (about 50 MHz) and extremely low temperatures (T < 1°K) in order to exclude the Van Vleck−de Krönig mechanism. This approach, as well as the simplicity of the spin system of the cyanide (the spin of the particles is $s = \frac{1}{2}$ and the EPR spectrum consists of a single absorption line) and the simplicity of the spectrum of ion pairs, enabled us to achieve at least partial success.

These two cases of strong (compared with the Zeeman mechanism) spin-spin exchange interactions in magnetically concentrated (DPPH) and magnetically dilute $[K_3(Co,Fe)(CN)_6]$ spin systems do not represent all possible deviations from the simple Van Vleck–de Krönig model [2, 3] of the spin-lattice interaction in paramagnetic crystals. For example, the case of a weak (compared with the Zeeman mechanism) exchange interaction in ruby has been considered by Al'tshuler [11]. This interaction explains, by analogy with the work of Waller [12], the concentration dependence of the relaxation rate in magnetically dilute ruby in which the mechanism involving ion pairs is effective [9].

Another possibility of generalizing the Van Vleck–de Krönig model is based on the fact that at very low temperatures the lattice cannot be regarded as an infinite thermostat with respect to spins. The oscillators interacting with the spin system must have frequencies corresponding to the resonance absorption lines because a long time is necessary for the establishment of a thermal equilibrium within the lattice. The number of such oscillators is small and their heat capacity is limited. The transfer of energy from the spins can heat these oscillators to a temperature higher than the temperature of the lattice as a whole. In this case, the surrounding space (for example, a liquid helium container, a paramagnetic salt in the magnetic cooling apparatus, etc.) can act as the thermostat.

We shall not consider the paramagnetic relaxation anomalies associated with the mechanisms considered in the last two paragraphs. We shall describe only the studies of the influence of strong (compared with the Zeeman mechanism) exchange interactions on the paramagnetic spin-lattice relaxation in crystals with high and low concentrations of magnetic particles.

The present paper consists of four chapters.

Chapter I describes experiments on the EPR in DPPH at and below liquid-helium temperatures. The results and method are compared with the work of other experimenters. Conclusions about the structure of DPPH are drawn from the temperature dependences of the Curie constant and the EPR line width. Some of these conclusions may be regarded as controversial.

Chapter II presents a brief review of the fundamentals of the theory of paramagnetic spin-spin and spin-lattice relaxation in very simple magnetic crystalline systems (consisting of spins $s = \frac{1}{2}$). Special attention is paid to the generalization of the classical Van Vleck–de Krönig relaxation theory by the inclusion of the spin-spin degrees of freedom in the relaxation mechanism. In this connection, a discussion is given of the influence of strong (compared with the Zeeman energy) electron exchange on the paramagnetic relaxation in magnetically concentrated systems. DPPH is considered as a typical representative of such systems. A phenomenological description is supplemented by the results which follow from the quantum theory of the probability of nonradiative transitions in crystals.

The results reviewed are used, to a considerable extent, as the theoretical basis of the material presented in Chapter III, which deals with the influence of strong exchange interactions on the spin-lattice relaxation in magnetically dilute systems. The review aims to provide the first comprehensive treatment of problems associated with the generalization of the Van Vleck–de Krönig model of the spin-lattice interaction.

The original work reported in Chapter II includes the results of measurements of the spin-lattice relaxation in DPPH carried out by the method of continuous saturation of the 42 MHz absorption line at liquid helium temperatures. This chapter also includes a discussion of similar results obtained by American workers at a frequency of about 10^4 MHz. These experimental results are reviewed in Chapter II on the basis of a theory of relaxation processes involving exchange interactions, which is generalized to the case of a finite specific heat of the crystal lattice (this case is important at low temperatures).

Chapter III reports the results of an experimental investigation of the influence of exchange-coupled Fe^{3+} ions (ion pairs) on the spin-lattice interaction of Fe^{3+} ions in magnetically dilute crystals of $K_3(Co,Fe)(CN)_6$ crystals with Fe^{3+} concentrations of $f \approx 10^{-3}$. As pointed out already, the special features of these experiments were the low temperature (T < 1°K) and the low frequency of the EPR absorption, which made it possible to exclude the Van Vleck–de Krönig spin-lattice interaction mechanism. This chapter also includes a detailed discussion of the spin-lattice relaxation in $K_3(Co,Fe)(CN)_6$, a theoretical calculation of the probabilities of nonradiative transitions involving ion pairs, and a comparison of the theoretical and experimental results.

Chapters II and III deal quite fully with the special properties of the spin-lattice relaxation in crystals, which are associated with electron exchange within the paramagnetic particle system.

Chapter IV describes the apparatus used in EPR experiments at extremely low temperatures (T < 1°K). The layout of the apparatus is considered, the construction of some units is discussed, the special aspects of the technique used at T < 1°K are analyzed, and some of the measuring circuits and their characteristics are detailed.

CHAPTER I

Electron Paramagnetic Resonance in α,α-Diphenyl-β-Picrylhydrazyl Free Radicals at Liquid-Helium and Lower Temperatures

§1. Review of the Results of Investigations of the Antiferromagnetic Exchange Interaction in Molecular Crystals of Chemically Stable Organic Free Radicals

Electron paramagnetic resonance in free radicals was first observed in 1947 by B. M. Kozyrev and S. G. Salikhov [13]. Their work stimulated intensive studies of organic substances and chemical reactions by the EPR method. The simplest organic radicals are those which form chemically stable molecular crystals. The best known of them is α,α-diphenyl-β-picrylhydrazyl (DPPH), which was first investigated in 1950 [14, 15].

A characteristic property of some organic free radicals is the small width of the EPR line [16, 17], which is 1-10 Oe at temperatures T = 300-77°K. According to the theory of the EPR line width, this indicates the existence of the following isotropic exchange interaction:

$$-\tfrac{1}{2} J (1 + 4\mathbf{s}_1 \mathbf{s}_2)$$

between unpaired electrons with the constant $J \approx 1\text{-}10°K$. The sign of J for DPPH and some other free radicals indicates an antiferromagnetic exchange interaction in these substances. This is in agreement with earlier measurements of the static magnetic susceptibility of some chemically stable organic free radicals [18-20].

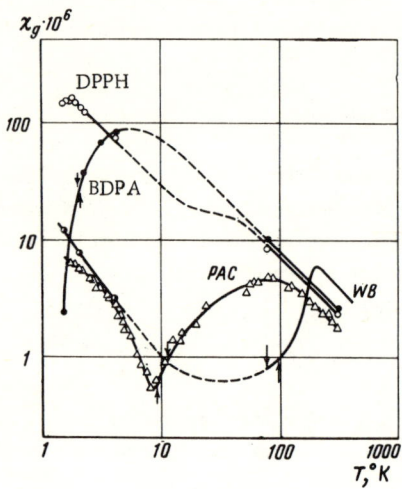

Fig. 1. Temperature dependence of the static magnetic susceptibility of free radicals. The experimental points indicate the intensity of the Larmor resonance at 25 MHz (at 10 kMHz for WB). The results for WB are taken from the measurements of the static susceptibility reported by N. Elliott and M. Wolfsberg, Phys. Rev., 91:435 (1953). The curve for DPPH is extrapolated to 4-77°K in agreement with [21]. The arrows below the curves indicate the temperatures at which resonance with g = 4 is observed. The arrows above the curves correspond to temperatures at which the g = 2 line width passes through a maximum.

This information was used in 1953-1954 in attempts to detect the paramagnetic-antiferromagnetic transition in fine DPPH powders during cooling to liquid helium temperatures. However, these attempts were not successful. Singer and Spencer [21] investigated the resonance absorption in DPPH at 25 MHz in the temperature range T = 295-2.5°K and showed that — in this range of temperatures — the temperature dependence of the static magnetic susceptibility had no maximum, which could indicate a transition to the antiferromagnetic state. However, they found that the line width increased sevenfold when a sample was cooled in the range T = 77-2.5°K. Other workers [22] reported in 1954 an investigation of the ~7 cm absorption line of DPPH at low temperatures down to 1.6°K. They found no appreciable shift of the absorption line. Measurements of the static magnetic susceptibility, $\chi_0 = \text{const}/(T - \Theta)$, carried out by these workers in fields H = 3.5-18 kOe in the temperature range T = 20-1.5°K yielded the Curie constant $\Theta = -0.1°K$. It was suggested in [22] that the transition to the antiferromagnetic state may occur below 1.5°K. All the experiments mentioned so far were carried out on fine powders of DPPH.

The negative results of the search for the transition to an ordered magnetic state, reported in [21, 22], and the interest in the antiferromagnetic exchange interaction in crystals of organic free radicals stimulated further, more detailed, investigations of the magnetic properties of DPPH and some other free radicals at low temperatures. In 1960-1962 a group of workers at Stanford University in the USA carried out a systematic EPR study of the properties of fine powders of four free radicals at low temperatures T = 300-1.5°K: 1) α,α-diphenyl-β-picrylhydrazyl (DPPH); 2) picryl-n-aminocarbazyl (PAC); 3) 1-3-bisdiphenylene-2-phenylallyl (BDPA); 4) Wurster's Blue perchlorate (WB) [23-28]. The results of their experiments are

TABLE 1.

Radical	Molecular weight	Assumed magnetic susceptibility · 10^6 (per mole)	Curie constant · 10^3 (°K/g)	Curie point, °K	Concentration of radical
PAC	392	−191	0.340	−51±7	36
WB	264	−129	1.33	−36±5	94
DPPH (C_6H_6)	472	−252	0.463	−25±3	58
DPPH (CS_2)	394	−197	0.876	−22±3	92
DPPH (CCl_4)	394	−197	0.817	−22±3	86
BDPA	496	−316	0.725	−6±4	96

TABLE 2

Free radical	Θ_H	Θ_L, °K	T_m, °K	$T_{\Delta H}$, °K	c
WB	-39 ± 15	0.6 ± 0.1	180	78	0.95
PAC	-130 ± 10	2.1 ± 0.1	75	12	
BDPA	-2.2 ± 1.0		6	1.8	0.96
DPPH	-15 ± 10	-0.4 ± 0.3	2		0.92

shown in Fig. 1 [25]. In addition to these experiments, the static magnetic susceptibilities of the same four radicals were determined at temperatures T = 300-77°K (Table 1) [29].

Three of these radicals — BDPA, PAC, and WB — behave similarly during cooling in the range T = 300-1.5°K; they exhibit the properties described below.

a) At high temperatures, the temperature dependence of the susceptibility of these three radicals, as well as that of DPPH, is given by the paramagnetic formula

$$\chi_H = \frac{c_H}{T - \Theta_H}$$

with a negative constant $\Theta_H \simeq -10°K$, indicating an antiferromagnetic exchange interaction between spins.

b) BDPA, PAC, and WB exhibit a broad susceptibility maximum when the temperature is lowered.

c) When the temperature is lowered still further, the susceptibility of PAC and WB passes through a minimum and rises again in accordance with the paramagnetic law

$$\chi_L = \frac{c_L}{T - \Theta_L},$$

where $c_L/c_H \approx 10^{-2}$. The constant is now positive, $\Theta_L \approx 1°K$, which corresponds to a ferromagnetic exchange interaction. The absolute values of the constants are listed in Table 2. These constants are found from the susceptibility [28]. The constants Θ_H and Θ_L are the high- and low-temperature Curie (Weiss) constants; T_m is the temperature at which the susceptibility reaches its maximum; $T_{\Delta H}$ is the temperature corresponding to the maximum value of the EPR line width; c is the magnetic-particle concentration.

d) A maximum of the resonance line width is observed in the temperature range corresponding to the falling parts of the temperature dependences of the susceptibilities of DPA, PAC, and WB.

e) An absorption line with a g factor of four is observed in the same temperature for these three free radicals.

A model which explains qualitatively these observations is given in [28]. This paper discusses a spin system of a concentrated solid solution, i.e., it is assumed that the magnetic moments are distributed at random throughout the host crystal lattice. Closely spaced magnetic particles are coupled by the antiferromagnetic exchange interaction. When the temperature is lowered, the temperature dependence of the static magnetic susceptibility of such a system exhibits a broad maximum, a minimum, and a rise in accordance with the paramagnetic law [30, 28]. A broad maximum in the $\chi(T)$ curve does not indicate a transition to the antiferromagnetic state but is related only to the short-range magnetic order. The long-range magnetic order and, consequently, the Néel transition temperature $T_N > 0$ are exhibited by the model considered if

$$f(z-1) > 1,$$

where f is the concentration of magnetic particles and z is the number of nearest neighbors of a given magnetic particle. If the Néel temperature T_N is sufficiently low compared with the temperature corresponding to the susceptibility maximum, it is difficult to infer a phase transition from the temperature dependence of the susceptibility. A clearer indication of the antiferromagnetic ordering in the model described [30, 28] is a sharp peak in the temperature dependence of the specific heat at the phase transition point. Such a peak is reported in [28] for BDPA at 1.8°K. This is in good agreement with the position of a maximum of the resonance line width of BDPA at t = 1.8°K (such a maximum is expected at T_N [31]). The line having g = 4, observed at low temperatures, is attributed to an increase in the line width and the partial lifting of the forbiddenness of the transition $\Delta m = 2$ (m is the magnetic quantum number, which is the eigenvalue of the operator representing the projection of spin onto a direction of the static magnetic field).

The studies carried out at Stanford University indicate that DPPH behaves differently from the other free radicals studied. It does not exhibit the properties characteristic of the model considered, which describes well the behavior of the other three radicals. DPPH exhibits no susceptibility maximum, although according to the Stanford group, such a maximum should be found near 1.8°K. In the temperature range T = 77–4.2°K, DPPH shows a transition from the law $\chi_H = c_H/(T - \Theta_H)$ with $\Theta_H = -15 \pm 10$°K to the law $\chi_L = c_L/(T - \Theta_L)$ with $\Theta_L = 0.4 \pm 0.3$°K (Table 2), and $c_L/c_H \approx 0.5$. A similar value of Θ_L follows also from measurements of the specific heat of DPPH at liquid helium temperatures [24]: $J = \Theta_L = -0.78$°K. This equation applies when each unpaired electron is exchange-coupled to two nearest neighbors (z = 2). The EPR line width increases during cooling from liquid-nitrogen to liquid-helium temperatures [21]. The line having g = 4 is not observed. The high-temperature Curie (Weiss) constants, Θ_H, of DPPH, BDPA, PAC, and WB are confirmed by measurements of the static susceptibility at T = 300–77°K [29] (Table 1).

In view of the anomalous behavior of DPPH, compared with the other three radicals investigated, the Stanford group extended their measurements on DPPH to temperatures below 1°K [28].

The experiments on DPPH carried out by the present author in 1961–1962 [32, 33] had the same purpose and were also performed in the temperature range T < 1°K. These experiments were planned and carried out without knowledge of the work of the Stanford group but were based on earlier results [18–22]. The experiments described here were stimulated also by measurements of the static magnetic susceptibility made (for the present author) in January, 1960 by Prof. J. Cohen at the request of Prof. M. Soutif of the Fourier Institute in France. These measurements were carried out on DPPH powder at temperatures T = 1.6–20°K in a

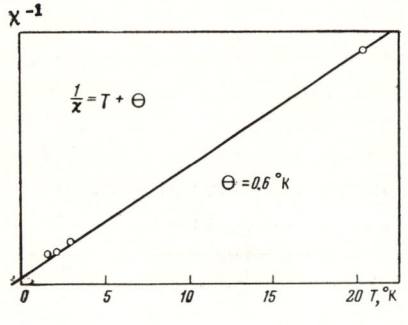

Fig. 2. Temperature dependence of χ^{-1} of a fine DPPH powder (J. Cohen's measurements.)

TABLE 3

Sample	$c \cdot 10^6$, cgs emu	% of radical	Θ, °K	$c_L \cdot 10^6$	Θ_L
I	743	78	−26	380	−0.4
II	919	97	−23	477	−0.4
III	919	97	−24	462	−0.4

magnetic field of 1000–1500 Oe. The results are shown in Fig. 2. The value obtained for $\Theta_L =$ −0.6°K agreed with the results given in Table 2. A similar result [24], $J = \Theta_L = -0.78$°K, was obtained for DPPH powder crystallized from benzene; the author became aware of it through Griffiths' paper [34]. These values of the Curie (Weiss) constant of DPPH as well as the results obtained by the present author and Prokhorov [32] were recently confirmed by measurements of the static susceptibility carried out at T = 1.2–294°K [35]. The results of these measurements are given in Table 3.

Samples I and III were crystallized from chloroform and sample II from benzene. Sample I was prepared several years earlier than samples II and III.

We shall consider in detail the results on free radicals obtained at Stanford because they are of direct relevance to the experiments on DPPH reported in the present paper. The comparison of our results with those of Stanford will be given later.

§2. Experimental Results

In later sections of the present chapter we shall describe investigations of DPPH at low temperatures carried out in order to detect a transition from the paramagnetic to the antiferromagnetic state.

A schematic representation of the DPPH molecule is given in Fig. 3. The magnetic properties of this radical are due to the presence in each molecule of one unpaired electron and are described by the following spin Hamiltonian

$$\hat{\mathcal{H}} = \sum_{i>k} J_{ik} \hat{S}_i \hat{S}_k + g\beta H \sum_i \hat{S}_{zi} + \hat{\mathcal{H}}_{\text{dip}}. \tag{1.1}$$

The interactions are given in this Hamiltonian in decreasing order of their magnitude. The first term represents the isotropic exchange interaction of unpaired electrons; the second is the Zeeman energy which, in the magnetic fields used in our experiments, is less than the exchange energy; the third term is the dipole spin-spin interaction. The z axis is the direction of a static magnetic field H; β is the Bohr magneton; the spectroscopic splitting factor is g = 2.0036 ± 0.0002 at 300°K [14, 15]; the spin is s = ½.

The EPR spectrum of DPPH consists of a single absorption line. In our experiments the spectrum was determined at 42 MHz in the temperature range from 4.2 to 0.1°K. In contrast to

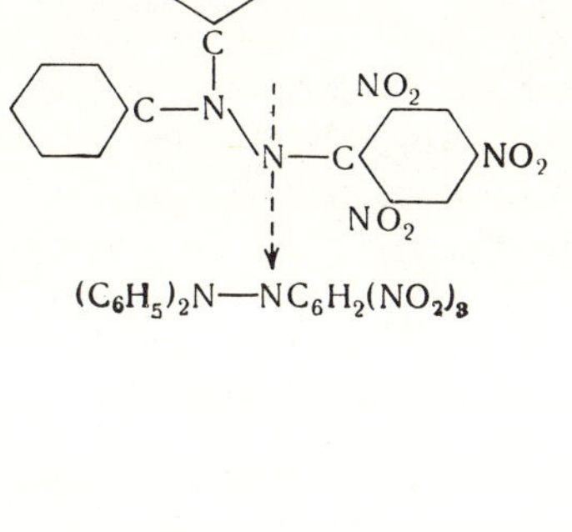

Fig. 3. Schematic representation of a molecule of DPPH.

$(C_6H_5)_2N-NC_6H_2(NO_2)_3$

the investigations described in §1, our measurements were carried out not only on fine powders but also on large single crystals of the radical.

The integral intensity of the EPR line decreased to zero at temperatures below 1°K. This disappearance of the EPR signal was due to changes in the energy spectrum of the system of magnetic moments of the unpaired electrons, which were caused by a transition of these electrons to an ordered (antiferromagnetic) state because of the presence of the exchange interaction in the system. The transition could be observed only when the Zeeman interaction was weak compared with the exchange interaction. This was why a low EPR frequency (42 MHz) was used in the present investigation.

The experimental dependences [32] are presented in Figs. 4 and 5. They were obtained using three samples of DPPH crystallized from a benzene solution: two of them were single crystals of 25 and 13.7 mg weight and the third sample was a powder of 50 mg weight. These crystals had a triclinic structure. The larger crystal was 2 mm thick and its maximum linear dimension was 5 mm. Cooling to 0.1°K was carried out by adiabatic demagnetization of iron-ammonium alum $FeNH_4(SO_4)_2 \cdot 12H_2O$. A reliable optical contact was established between a sample and the alum by a heat conductor, one end of which penetrated into the alum and the other of which was embedded in a frozen drop of glycerin which surrounded the sample. The temperature was controlled by the magnetic susceptibility of the paramagnetic salt with an accuracy of not less than 10%. The resonance absorption was detected with an autodyne oscillator operating at a frequency of 42 MHz, whose circuit included the substance under investigation. The static magnetic field used to observe the resonance was modulated at a frequency of 0.5 Hz. The oscillation amplitude, the linearity of the spectrometer, and the absence of the saturation effect in the sample were checked. The saturation effect was detected with a calibrator (reference) circuit suggested by Watkins [36]. The signal/noise ratio for the larger crystal was about 100. The absorption line was recorded by photographing the screen of an oscilloscope.

The integral intensity of the line s was defined as the product of the line width ΔH and the amplitude I at the absorption curve maximum and it was expressed in relative units. This was correct only if the line profile was independent of temperature because $s = K(I \cdot \Delta H)$ where $K \approx 1$ is a coefficient which depends on the line profile. This coefficient is 0.94 and 0.64 for the Gaussian and Lorentzian profiles, respectively; the actual profiles of the absorption lines are usually intermediate between the Gaussian and Lorentzian curves. The absolute measurements of the line amplitude I were carried out with the Watkins calibrator circuit referred to earlier.

Chapter IV gives more details of the apparatus used and of the experimental conditions in the spin resonance experiments at very low temperatures.

§3. Temperature Dependence of the Integral Intensity of the Resonance Absorption Line

It is mentioned in §2 that a common property of the investigated samples (fine DPPH powder and large single crystals) is the disappearance, in a narrow range of temperatures, of the EPR line if the temperature is made sufficiently low. However, there are considerable differences between the experimental dependences (Fig. 4) for the larger single crystal and the fine powder. Therefore, in discussing the results we shall mention the sample to which they apply.

Let us consider first the experimental data for the larger DPPH single crystal (curve 1 in Fig. 4). They can be summarized as follows. The integral intensity of the EPR line of this crystal obeys the law $s = \text{const}/(T - \Theta)$ with $\Theta = -0.35°K$ right down to $T \approx 0.35°K$; below this temperature the intensity deviates from this law, decreases rapidly, and vanishes at $T \approx 0.18°K$. The main change in the integral intensity occurs in the range $T = 0.2$–$0.3°K$. The temperature

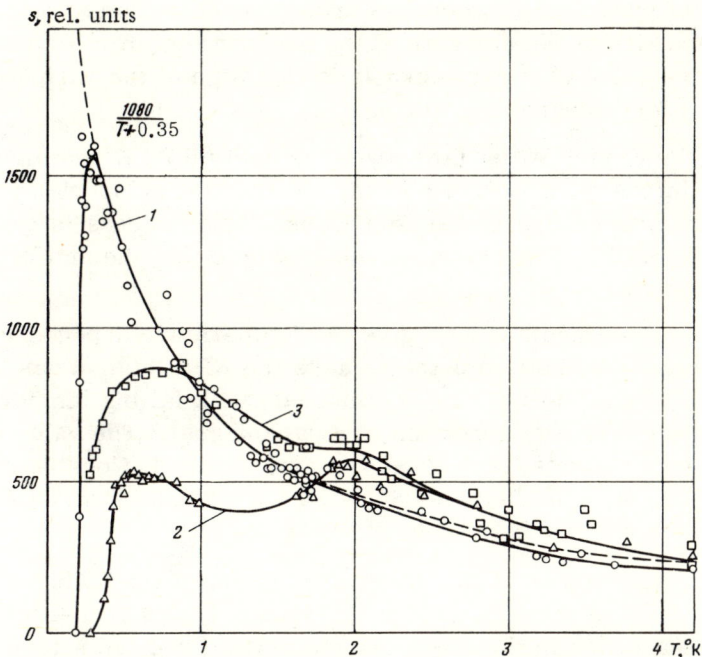

Fig. 4. Temperature dependences of the integral intensity s of the EPR line for three samples of DPPH: 1) large single crystal of 25 mg weight; 2) crystalline powder of 50 mg weight; 3) small single crystal of 13.7 mg weight (curves 2 and 3 are normalized to the weight of 25 mg).

dependence of the line intensity is not affected by the orientation of the crystal in the static magnetic field. The position of the maximum of the absorption curve is independent of temperature to within 10%.

In the paramagnetic region the static susceptibility of the radical is proportional to the area under the EPR line. According to the Hamiltonian (1.1), the imaginary component of the complex magnetic susceptibility $\chi''(\nu)$ has a single maximum near the frequency $\nu_0 = g\beta H/h$. Consequently, the integral intensity of the paramagnetic resonance absorption line is

$$s = \int_0^\infty \chi''(\nu)\, d\nu.$$

It follows from the Kramers-Krönig relationship

$$\chi_0 = \chi'(0) = \frac{2}{\pi} \int_0^\infty \frac{\chi''(\nu)}{\nu}\, d\nu$$

that if the resonance line width is small, $\Delta\nu \ll \nu_0$, we obtain

$$\chi_0 \approx \frac{2s}{\pi \nu_0}.$$

Thus, the experimental dependence $s = \text{const}/(T - \Theta)$ is simply the Curie-Weiss law for the investigated sample. According to the usual terminology, the temperature $\Theta = -0.35°K$ is the limit of the existence of paramagnetism and is known as the paramagnetic Curie or Néel point. The value of Θ is a measure of the strength of the exchange interaction in a system of unpaired electrons and a negative value of Θ corresponds to that sign of the exchange integral which is associated with the antiferromagnetic ordering.

Deviation from the Curie–Weiss law, observed below 0.35°K and manifested by a fall of the EPR integral line intensity, indicates a change in the energy spectrum of the radical such that the number of paramagnetic particles decreases. This process is most rapid in the temperature range T = 0.2-0.3°K. Therefore, we may assume that the antiferromagnetic Néel point lies in this temperature range.

We note that our experiments cannot give the temperature dependence of the static magnetic susceptibility near the transition point because the EPR method can be used to determine only the magnetic parameters (decreasing continuously with falling temperature) of that part of the sample which has the energy spectrum of a paramagnet. The value of the constant $\Theta = -0.35°K$, found in our experiments, is in good agreement with J. Cohen's result for the powder, which is $\Theta = -0.6°K$ (Fig. 2), with the value $\Theta_L = -0.4 \pm 0.3°K$ [28], and with later measurements on the powder [35], which yielded $\Theta_L = 0.4°K$.

The temperature dependence of the integral intensity of the EPR line of the fine powder sample also indicates a phase transition (Fig. 4). However, this transition begins at a higher temperature than the transition in the larger single crystal and the nature of the transition is more complex. The curve obtained for the smaller single crystal (curve 3 in Fig. 4) occupies an intermediate position between the curves for the larger crystal and the fine powder.

The two maxima of curve 2 of Fig. 4 indicate that the spin system of the fine powder consists of two subsystems with a different exchange interaction in each subsystem. The interaction constant of one of these subsystems is approximately $J_1 \approx 2°K$ and the constant for the other subsystems is $J_2 \approx 0.4-0.5°K$. According to curve 2, the number of particles in the first subsystem is several times higher than in the second. Therefore, at higher temperatures (T > >2°K), the properties of the spin system of the DPPH powder are governed by particles of the first subsystem. However, in the larger single crystal the exchange interaction in the spin system is represented by only one exchange integral, $J \approx 0.35°K$, and the total density of unpaired electrons is evidently somewhat lower than in the fine powder.

These conclusions follow from an analysis of the experimental data. An additional confirmation is provided by the explanation of the observed difference between the line widths of the powder and the larger single crystal (Fig. 5). The narrower line of the powder corresponds to a stronger exchange interaction that that in the crystal and, consequently, it indicates a more pronounced exchange narrowing of the absorption line (§ 4).

However, the experimental data are insufficient to determine the cause of the observed difference between the larger single crystal and the powder. The nature of the results for the smaller single crystal of 13.7 mg weight (curves denoted by 3 in Figs. 4 and 5) suggests that the difference between the data for the larger crystal and the fine powder, and the difference between the exchange interactions in these samples are both due to the different crystal structures of the free radical in the fine-powder and large single-crystal forms.

It is interesting to note that our results for the fine DPPH powder obtained at T > 0.6°K are similar to the results of the Stanford group (§ 1) obtained for the three radicals BDPA, WB, and PAC: there is a broad maximum followed by a minimum in the temperature dependence of the DPPH line intensity and the temperature dependence of the line width has a maximum. This analogy suggests that the difference between the results obtained for the fine powder and the larger single crystal of DPPH should also be observed for the BDPA, WB, and PAC radicals. It would be interesting to check experimentally this hypothesis. The model of a magnetically dilute spin system given in [30, 28] (§ 1) must be refined before it can be used to interpret the

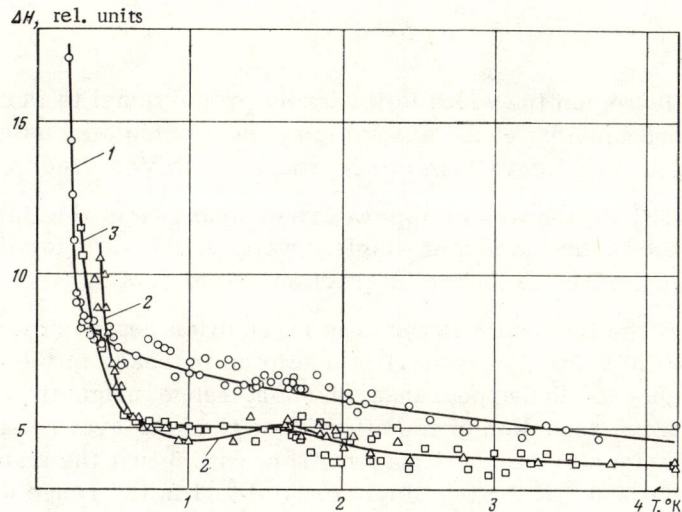

Fig. 5. Temperature dependence of the resonance line width ΔH for the three samples described in detail in Fig. 4.

properties of DPPH in the fine powder form; this model must include the structural differences between the powder and the larger single crystal, which are manifested by the experimentally observed changes in the exchange interaction. The final determination of the cause of the difference between the properties of large single crystals and fine powders requires further experimental work.

In conclusion, we must mention that the EPR line of DPPH does not disappear if measurements are carried out on a fine powder at T = 0.25°K at a wavelength of about 3 cm [37]. The "hysteresis" observed in these experiments is attributed [37] to the dependence of the demagnetization field on the heating of a sample by the hf field. The phase transition is not observed because, in contrast to our experiments, the Zeeman splitting in the experiments reported in [37] is comparable with the exchange interaction constant of the radical.

§4. Temperature Dependence of the Resonance Line Width

The experimental results on the magnitude of the interaction in DPPH and on the phase transition point, discussed so far, can be supported and supplemented by an analysis of the experimentally obtained temperature dependence of the EPR line width.

The temperature dependences of the line width for the larger single crystal and the fine powder (see preceding section) are represented by curves 1 and 3 in Fig. 5. At T = 300°K, the EPR line width for the powder is $\Delta H = 1.75 \pm 0.1$ Oe, but at T = 77°K the width is $\Delta H = 1.5 \pm 0.1$ Oe. The curves in Fig. 5 have two easily distinguishable regions: a) the paramagnetic region at T > 0.4-0.6°K, characterized by a weak temperature dependence of the line width, and b) the phase transition region, characterized by a rapid rise of the width at T < 0.4-0.6°K.

In the paramagnetic region the EPR line width of DPPH is described by the theory of exchange narrowing of the EPR lines [17, 38-41]. The exchange narrowing effect is a reduction of the dipole spin-spin width due to the averaging of the internal local magnetic field acting on a given paramagnetic particle (the averaging is due to electron exchange between the paramagnetic particles). There is no magnetic order in the paramagnetic region and the electron exchange process is random. In this case the correlation time τ_k of the exchange motion is equal to the period of electron exchange $\tau_k \approx h/J$. The EPR line width given by the theory [17, 39] can be expressed in terms of the correlation time $\tau_k = 1/\nu$, by the relationship

$$\Delta \nu = \frac{2 \langle \Delta \nu^2 \rangle}{\nu_k} \qquad (1.2)$$

and in the paramagnetic region the width is inversely proportional to the constant J. The quantity $\langle \Delta \nu^2 \rangle$ is the second moment of the absorption curve, calculated using the Hamiltonian of the magnetic dipole spin-spin interaction in accordance with Van Vleck's theory [38].

According to Eq. (1.2), the weak temperature dependence of the line width in the paramagnetic region (T = 4.2-0.4°K for the larger single crystal and 77-2°K for the fine powder) can be explained by a weakening of the exchange interaction due to cooling.

The rapid rise of the line width in the phase transition region (T < 0.4°K for the larger single crystal and T < 0.6°K for the powder) is due to an increase in the correlation time of the exchange motion because of the appearance of short-range magnetic order in the spin system and the corresponding reduction of the effectiveness of the averaging of the internal magnetic field by the exchange motion. It is evident from Fig. 5 that the transition is most rapid in the temperature range 0.2-0.3°K for the single crystal and in the range 0.4-0.5°K for the powder. This is in agreement with the conclusions drawn from the temperature dependence of the integral intensity of the resonance absorption line.

We shall consider in more detail [33] the change in the exchange interaction caused by cooling the DPPH powder since, in addition to the experimental data on the line width of the powder, we know also the temperature dependence (in a wide range of temperatures) of the Curie constant, which is related directly to the exchange integral J:

$$k\Theta = \frac{2}{3} z J s (s+1).$$

For a system of particles with spin $s = \frac{1}{2}$, we obtain

$$\Theta = zJ / 2k. \qquad (1.3)$$

Here z is the number of unpaired electrons in the crystal lattice of nearest neighbors of a given particle; k is the Boltzmann constant. An analysis of this problem can be used to draw some conclusions about the structure of the spin system of the DPPH powder.

The conclusion regarding the weakening of the exchange interaction in the DPPH powder at T = 77-2°K follows from the increase in the EPR line width and from the discrepancy (§ 1) between the Curie constant Θ_H, found from measurements of the static magnetic susceptibility of the powder at "high" temperatures T = 300-77°K, and the value of the Curie temperature of the powder Θ_L, found from measurements at low temperatures (below 20°K). In fact, $\Theta_H = -(10-20)$°K according to [18-20]; $\Theta_L = -0.6$°K according to J. Cohen (Fig. 2); and $\Theta_L = -0.1$°K according to Gerritsen et al. [22]. A similar conclusion that Θ and, consequently, J decrease at low temperatures is reached in [28] on the basis of the experimental data reported in [24-27, 29]. Recently, this result has been confirmed by measurements of the static magnetic susceptibility of the DPPH powder carried out in the range 300-1.2°K [35].

The reduction of the magnitude of the exchange integral at low temperatures is evidently due to a change in the crystal structure. One of the indications of such a change is an increase in the g-factor anisotropy at helium temperatures [1]. The drop in J in the temperature range 77-2°K is described quantitatively by the relationships

$$\frac{\Theta_{77°K}}{\Theta_{2.0°K}} = \frac{J_{77°K}}{J_{2.0°K}} = \frac{\exp(-\lambda d_{77°K})}{\exp(-\lambda d_{2.0°K})} \approx 60, \qquad (1.4)$$

$$\frac{\Delta H_{2.0°K}}{\Delta H_{77°K}} = \frac{J_{77°K}}{J_{2.0°K}} \left(\frac{d_{77°K}}{d_{2.0°K}} \right)^6 \approx 3. \qquad (1.5)$$

These relationships are deduced from Eqs. (1.2) and (1.3) and from the experimental data on the Curie temperature Θ [35]. The values of the EPR line width are taken from the experimental data reported in the present paper. Equation (1.4) is obtained by assuming that the exchange integral is an exponential function of the distance. Moreover, both Eqs. (1.4) and (1.5) are derived on the assumption that each unpaired electron has two nearest neighbors (z = 2) separated by a distance d. In order to satisfy both these relationships, we must assume that the distance between the nearest spins increases when the temperature of the radical is lowered: $d_{2.0°K}/d_{77°K} = 1.65$. We note that $z \neq 6$, because in this case we would have to assume that the volume of the unit cell increases when the crystal is cooled in the range T = 77-2°K.

The result z = 2 is important. It shows that the spin system of the powder consists, at sufficiently high temperatures, of an array of spin chains such that the exchange interaction within each chain is stronger than the interaction between different chains. The cooling of the radical to liquid helium temperature increases the distance between particles in a chain, reduces the exchange interaction within a chain, and can make this interaction comparable with the interaction between particles belonging to different chains. In this case, the spin system becomes of the "volume" type. Such a structure of the spin system of the DPPH powder does not contradict the experimentally observed temperature dependence of the resonance absorption intensity, which indicates (§ 3) the presence of two exchange interaction constants: J_1 and J_2.

On the other hand, the suggested structure is in good agreement with a rough model of the crystal structure of the radical (Fig. 6), which is based on the assumption of "close packing" of the molecules (the bond lengths are taken from [42, 43]). It follows from this lattice model that $(a/d) \approx (b/d) \gtrsim 2$, where d is the dimension of the cell along the N−N bond in the radical molecule. This confirms the validity of the assumption that z = 2 at "high" temperatures. The average size of the cell, obtained by dividing the volume of a crystal by the number of unpaired electrons, is approximately $8 \cdot 10^{-8}$ cm [34] and, therefore $a \approx b \approx 10^{-7}$ cm, $d_{77°K} \approx 5 \cdot 10^{-8}$ cm. This is in good agreement with the value $d_{77°K} \approx 3 \cdot 10^{-8}$ cm found from Eq. (1.4) for $\lambda \approx 2 \cdot 10^8$ cm^{-1} [34].

Conclusions

The main result of the reported experiments on free DPPH radicals is the observation of a transition from the paramagnetic to the ferromagnetic state at T < 1.0°K. This is the first time that such a transition has been observed in molecular crystals of organic free radicals. The discovery was made possible by using low temperatures and observing magnetic resonance at low frequencies.

It would be interesting to investigate experimentally the energy spectrum of the DPPH radical by the spin resonance method at temperatures below the transition point to the antifer-

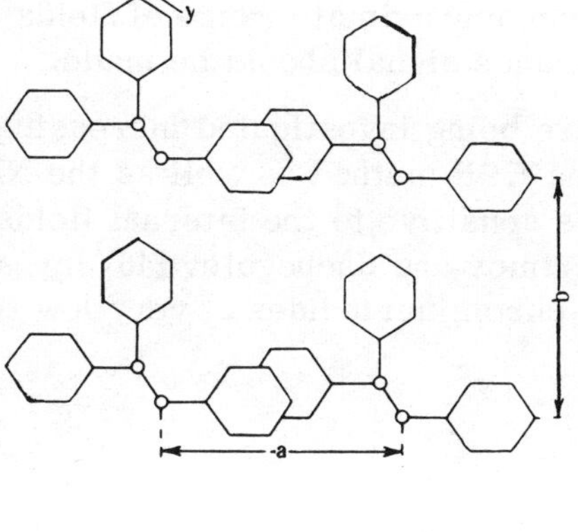

Fig. 6. Model of the crystal lattice of DPPH. The lattice consists of layers shifted somewhat relative to one another. Circles are used to represent nitrogen atoms.

romagnetic state. The fields and frequencies at which an antiferromagnetic resonance signal can be expected depend appreciably on the critical field of the antiferromagnet (the critical field is that value of an external static magnetic field above which the sublattice magnetizations are oriented in a plane perpendicular to the external field). According to [44, 45]

$$H_{cr} = \sqrt{\frac{2K}{\chi_\perp - \chi_\parallel}},$$

where K is the anisotropy energy constant of a substance in the case of axial anisotropy; $\chi_\parallel$ and $\chi_\perp$ are the magnetic susceptibilities of a substance for the parallel and perpendicular orientations of the static magnetic field with respect to the crystal axis. The resonance frequency formulas are simplest for crystals with axial anisotropy and we shall use them in our estimates. At low temperatures (much lower than the transition point), the critical field is

$$H_{cr} \approx \sqrt{\frac{2K}{\chi_\perp}} = \sqrt{2KA}, \tag{1.6}$$

where A is a constant in the molecular field theory given by $A = 4Jz/Ng^2\beta^2$ (z is the number of nearest neighbors of a given magnetic particle, N is the total number of magnetic particles in a system). The anisotropy energy may be related to the anisotropy of the crystal field or to the exchange interaction anisotropy. According to [44], the order of magnitude of K is approximately the same in these two cases (given respectively):

$$K \approx N\lambda(g-2), \tag{1.7a}$$

$$K \approx NJ(g-2)^2. \tag{1.7b}$$

Here λ is the spin-orbit interaction constant. Using Eq. (1.7b) and J = 0.4°K, z = 6, g−2 ≈ 5·10⁻³ [1], we find that $H_{cr} \approx$ 100 Oe. In external fields H weaker than the critical field, the resonance frequency ω is given by the ratio ($\gamma = g\beta/\hbar$) [44, 45]

$$\frac{\omega}{\gamma} = H_{cr} \pm H.$$

We can easily calculate that in order to observe resonance at 42 MHz, we require a field of about 85 Oe. However, it is found experimentally that at T = 0.1°K there are no lines at 42 MHz in fields up to 150 Oe. This may be due to the resonance line width being greater than the depth of modulation of the field (20-30 Oe) or to the critical field exceeding 100 Oe. Irrespective of whether the line is too wide, it is obvious that the field is too weak. In fact, it follows from $\lambda \approx \Delta(g-2)$, where Δ is the splitting of the energy levels in the crystal field, that the ratio of the expressions in Eqs. (1.7a) and (1.7b) is Δ/J. The order of this quantity for the DPPH radical is $\Delta/J \approx 100$ and, therefore, the calculated value of the critical field should be increased in accordance with Eq. (1.6) by an order of magnitude, i.e., $H_{cr} \approx$ 1000 Oe. These estimates are very rough. They simply indicate an approximate range of fields and frequencies in which a search for the antiferromagnetic resonance signal should be made.

The magnetic properties of free radicals are being investigated increasingly more thoroughly at low and very low temperatures using the EPR method as well as the NMR technique because the frequency of the nuclear resonance is sensitive to the internal fields in a substance. In addition to the radicals already mentioned, Karimov and Shchegolev [46] investigated the properties of dibenzene-chromium and ditoluene-chromium iodides at very low temperatures.

CHAPTER II

Influence of Exchange Interactions on Paramagnetic Relaxation in Magnetically Concentrated Systems

§1. Experimental Methods for the Investigation of Paramagnetic Relaxation

In electron paramagnetic resonance we observe a signal representing the resonance absorption of the energy from an external e.m. radiation field corresponding to quantum transitions between the energy levels of spins in a static magnetic field. Measurements of the rates of paramagnetic relaxation processes are usually based on determinations of two characteristics of the EPR signal which depend on the probability of relaxation transitions: the absorption line width and its intensity. Relaxation processes, tending to re-establish the thermal equilibrium disturbed by the external radiation, involve nonradiative quantum transitions. These transitions are due to the spin-spin and spin-lattice interactions and studies of these transitions are therefore a source of information on the internal interactions in crystals.

We can distinguish two types of nonradiative transition in experiments in which processes occurring in the Zeeman system of levels are being investigated. Transitions of the first type do not alter the Zeeman energy of spins and simply establish an internal thermal equilibrium in the Zeeman level system. Transitions of the second type involve changes in the Zeeman energy of spins. These transitions result in the transfer of the excitation energy from the Zeeman spin degrees of freedom to other spin degrees of freedom and to degrees of freedom corresponding to the thermal vibrations of the lattice.

Let us consider how the experimentally measured quantities (the line width and intensity) are related to the probabilities of the nonradiative transitions of the first ($1/\tau_z$) and second ($1/\tau$) type. Here, τ_z and τ represent the relaxation times of the corresponding transitions.

Both relaxation transitions contribute to the line width:

$$\Delta\nu \approx \left(\frac{1}{\tau_z} + \frac{1}{\tau}\right),$$

because they limit the lifetime of the Zeeman spin states. In the limiting cases, $\tau_z \ll \tau$ and $\tau \ll \tau_z$, the measurements of the line width give one of these relaxation times (τ_z or τ).

In the frequently encountered case $\tau_z \ll \tau$, the relaxation time τ is found from experiments involving saturation of the resonance absorption line which are based on the measurement of the integral line intensity. When the amplitude of the external hf radiation field H_{hf} is not too large, so that

$$\boldsymbol{\mu} \cdot \mathbf{H}_{hf} \ll h(\Delta\nu)$$

($\boldsymbol{\mu}$ is the magnetic moment of a particle), the states of the system of spins can be described by spin populations of single-particle Zeeman energy levels. The inequality governing the hf field amplitude implies that the interaction of particles with the radiation field is weak compared with the spin-spin interactions responsible (in this case) for the broadening of the resonance absorption line. For simplicity, we shall consider a system of particles with spin $s = \frac{1}{2}$. The EPR spectrum of such a system consists of a single absorption line. The standard equation for the difference of populations (proportional to the absorption intensity) of two Zeeman spin levels subjected to the radiation field is of the form

$$\frac{dx}{dt} = (1-x)w_{21} - (1+x)w_{12} - 2xW. \tag{2.1}$$

Here $x = (N_1 - N_2)/N$, where N_1 and N_2 are the populations of the Zeeman levels E_1 and E_2 ($E_2 > E_1$) of a magnetic particle; $N = N_1 + N_2$ is the total number of magnetic particles in the system; W is the probability of a transition between the levels E_1 and E_2 induced by the external radiation field; w_{12} and w_{21} are the probabilities of nonradiative transitions of individual particles in the system, where

$$w_{21} = w_{12} \exp \frac{E_2 - E_1}{kT}.$$

The equilibrium value of the difference between the populations (in the absence of the radiation field)

$$x_0 = \frac{w_{21} - w_{12}}{w_{12} + w_{21}}$$

can be used to rewrite Eq. (2.1) in the form

$$\frac{d(x - x_0)}{dt} = -(x - x_0)(w_{12} + w_{21}). \tag{2.2}$$

The solution of this equation shows that the difference between the populations after switching off the external radiation field tends exponentially to an equilibrium value

$$x(t) - x_0 = [x(0) - x_0] \exp\left(-\frac{t}{\tau}\right) \tag{2.3}$$

and the relaxation time τ of this process is defined by

$$\frac{1}{\tau} = w_{12} + w_{21}.$$

On the other hand, the solution under steady state conditions ($dx/dt = 0$) in the case of continuous saturation of the resonance transition can be found from Eq. (2.2):

$$\frac{x}{x_0} = \frac{1}{1 + 2W\tau}. \tag{2.4}$$

The quantity x/x_0 is called the saturation factor.

The dependences (2.3) and (2.4) are used as the basis of two different methods of measuring the relaxation time τ. The first method consists of measuring the time dependence of the difference of populations of two levels (or of the absorption signal intensity), when the system returns to the state of thermodynamic equilbrium with the lattice after the external radiation field is switched off. In the second method we measure the value of the steady-state difference between the populations at a fixed intensity of the applied hf field. The second method of investigating paramagnetic relaxation was used by the present author at low frequencies.

The next section describes briefly and usually qualitatively the results of phenomenological and quantum-mechanical theories of the paramagnetic relaxation in the simplest systems. The intention is to find the physical meaning of the relaxation times τ and τ_z and to understand the whole fairly complex relaxation process in a crystal, as found by the experimental methods just described.

§2. Brief Results of the Theory of Paramagnetic Relaxation in the Simplest Systems

The theory of paramagnetic relaxation in crystals is based on the Hamiltonian

$$\hat{\mathcal{H}} = \hat{\mathcal{H}}^S + \hat{\mathcal{H}}^L + \hat{\mathcal{H}}^{SL}, \tag{2.5}$$

where $\hat{\mathcal{H}}^S$ is the energy operator corresponding to the spin degrees of freedom; $\hat{\mathcal{H}}^L$ is the energy operator of the thermal vibrations of the lattice; $\hat{\mathcal{H}}^{SL}$ is the operator of the spin-lattice interaction of all the spin degrees of freedom which contribute to $\hat{\mathcal{H}}^S$.

The spin Hamiltonian of the simplest system of spins $s = \frac{1}{2}$

$$\hat{\mathcal{H}}^S = \hat{\mathcal{H}}^Z + \hat{\mathcal{H}}^{exch} + \hat{\mathcal{H}}^{dip} \tag{2.6}$$

consists of the Zeeman energy of spins

$$\hat{\mathcal{H}}^Z = \sum_i \beta \mathbf{H} g \cdot \mathbf{s}_i$$

(β is the Bohr magneton; $\mathbf{H}$ is the static magnetic field; $\mathbf{s}_i$ is the spin of an i-th particle) and the energy of the spin-spin interactions, including the isotropic exchange interaction

$$\hat{\mathcal{H}}^{exch} = -\sum_{i,k} \frac{J}{2}(1 + 4\hat{s}_i \hat{s}_k)$$

(J is the exchange integral) as well as the magnetic dipole-dipole interaction of spins

$$\hat{\mathcal{H}}^{dip} = \hat{\mathcal{H}}_1^{dip} + \hat{\mathcal{H}}_2^{dip}.$$

The matrix of the operator $\hat{\mathcal{H}}_1^{dip}$ is diagonal in the S_z representation, i.e., in the representation based on the projection of the total spin of the system onto the direction of the static magnetic field; in this representation the matrix of the operator $\hat{\mathcal{H}}_2^{dip}$ has only nondiagonal matrix elements corresponding to a change of M (the eigenvalue of the operator $\hat{S}_z$) by an amount $|\Delta M| = 1$ or 2. The explicit form of the operators $\hat{\mathcal{H}}_1^{dip}$ and $\hat{\mathcal{H}}_2^{dip}$ is fairly cumbersome and, therefore, we shall not consider it further. The full expressions are well known and they are given, for example, in [17, 38, 41, 47].

The energy operators $\hat{\mathcal{H}}^Z$ and $(\hat{\mathcal{H}}^{exch} + \hat{\mathcal{H}}_1^{dip})$ commute. Therefore, the corresponding Zeeman and spin-spin degrees of freedom can be regarded as independent. Van Vleck [48] has given the physical meaning of the treatment of various degrees of freedom proposed by Bloembergen and Wang [4]. The problem is that the Zeeman and spin-spin energies are characterized by different quantum numbers: the former by the quantum number M and the latter by the total spin of the crystal S. Since there are many particles in a paramagnetic crystal, each value of M corresponds to a large number of values of S, and conversely. Consequently, the Zeeman and spin-spin energies can in fact vary independently. They are related only through the interaction $\hat{\mathcal{H}}_2^{dip}$.

The theory of paramagnetic relaxation deals with the processes of distribution between various spin degrees of freedom, of the energy which spins absorb from the external radiation field, and with the transfer of excitation energy to the degrees of freedom corresponding to the thermal vibrations of the lattice.

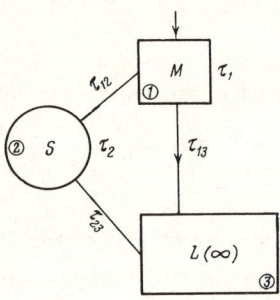

Fig. 7. Schematic representation of the paramagnetic relaxation mechanism in crystals. M, S, and L and numbers 1, 2, and 3 denote, respectively, the systems of the Zeeman, spin-spin, and lattice degrees of freedom.

The classification of the degrees of freedom suggested here allows us to represent the relaxation mechanism as shown schematically in Fig. 7. In this figure, the lattice is represented as a spin thermostat because, right down to very low temperatures, its specific heat is large compared with the specific heat of the spin systems. The transfer of energy from the spins to the lattice is characterized by five relaxation times corresponding to nonradiative transitions caused by internal interactions in each of the systems of the degrees of freedom (τ_1 and τ_2), and to coupling between the degrees of freedom of different systems (τ_{12}, τ_{23}, τ_{13}).

In general, the paramagnetic relaxation mechanism is complex. Therefore, we shall consider two limiting cases of the schematic representation in Fig. 7. In the first case, the energy of excitation of the spins is transformed into the energy of thermal motion by the spin-lattice interaction of the Zeeman degrees of freedom with the lattice. In the second case, the energy is transferred first from the Zeeman system to the spin-spin degrees of freedom and then to the lattice.

In the first case,

$$\mathcal{H}^z \gg \mathcal{H}^{\text{exch}}, \mathcal{H}^{\text{dip}}, \tag{2.7}$$

the establishment of equilibrium between the Zeeman and spin-spin degrees of freedom is difficult because of a large difference between the characteristic frequencies of the two systems, so that we can neglect the flow of energy from the Zeeman system through the spin-spin system to the lattice. The inequality of Eq. (2.7) is satisfied at high frequencies and in strong magnetic field by samples with low concentrations of magnetic particles.

In the second case, the spin-spin interactions must not be small compared with the Zeeman interactions. We shall assume that the following inequality applies:

$$\mathcal{H}^{\text{exch}} \gtrsim \mathcal{H}^z > \mathcal{H}^{\text{dip}}. \tag{2.8}$$

A good example of this case is the free radical DPPH in which the first mechanism is inactive because of the weak spin-orbit interaction [4].

The classical theory of paramagnetic relaxation developed by de Krönig [2] and Van Vleck [3, 38] corresponds to the first of the limiting cases discussed in the preceding paragraphs. The generalization of this theory to include the spin-spin degrees of freedom was made by Bloembergen et al. [4, 47]. These papers provide a complete description of paramagnetic relaxation in the simplest systems described by the Hamiltonian of Eqs. (2.5) and (2.6).

We shall now consider in detail the two limiting cases of the theory of paramagnetic relaxation. We shall discuss first a phenomenological description of the relaxation process in order to find the relationship between the experimentally measured relaxation times τ_z, τ (§ 1), and the probabilities of nonradiative transitions characterized by the time constants τ_1, τ_2, τ_{12}, τ_{23}, and τ_{13} (Fig. 7). The range of validity of the various forms of the phenomenological theory can then be found from the microscopic theory of probabilties of nonradiative transitions in a crystal, based on the Hamiltonian of Eqs. (2.5) and (2.6).

In the first limiting case, corresponding to the inequality (2.7), the paramagnetic relaxation is characterized by two time constants: the time τ_1 required for the establishment of an equilibrium within the Zeeman system, and the time τ_{13} for the establishment of an equilibrium between the Zeeman degrees of freedom and the lattice. It is obvious that the times τ_1 and τ_{13} correspond directly to the times τ_z and τ introduced in § 1.

If $\tau_1 \ll \tau_{13}$ and if the equilibrium within the Zeeman system is established more rapidly than that between the lattice and the Zeeman system, the state of the Zeeman system can be represented by a temperature T_1 (sometimes called the "spin temperature"), which is higher than the lattice temperature T_0. In this case, the experiments on continuous saturation are described by the equation

$$P = \frac{E_1(T_1) - E_1(T_0)}{\tau_{13}}. \tag{2.9}$$

If the Zeeman splitting is small, $\delta \ll kT_0$, we obtain the following relationships for the energy of the Zeeman system $E_1(T)$ and for the absorbed power P:

$$E_1 = \text{const} - \frac{A_1}{kT_1} \quad \text{for} \quad A_1 = \frac{1}{4} N \delta^2,$$

$$P = \frac{2WA_1}{kT_1}.$$

Here W is the probability of an induced transition between the Zeeman sublevels and N is the number of magnetic particles in a sample. The solution of Eq. (2.9),

$$\frac{T_0}{T_1} = \frac{1}{1 + 2W\tau_{13}}, \tag{2.10}$$

gives the value of the experimentally determined saturation factor [Eq. (2.4) in § 1]. The relationships (2.9) and (2.10) are identical with Eqs. (2.1) and (2.4).

If $\tau_1 \gg \tau_{13}$, the Zeeman system should be represented by a range of temperatures. In this case, equilibrium within the Zeeman system and between the Zeeman system and the lattice is established by spin-lattice nonradiative transitions in a time interval τ_{13}. In other words, the Zeeman system and the lattice should be regarded as a single thermodynamic system for which the time of establishment of the internal equilibrium is τ_{13}.

A quantum-mechanical theory of the spin-spin (time τ_1) and spin-lattice (time τ_{13}) relaxation is given in [3, 17, 38, 41].

De Krönig and Van Vleck [2, 3] demonstrated that the alternating electric field generated by the lattice vibrations acts on isolated spins because of the spin-orbit coupling. The modern theory of the spin-lattice interaction in paramagnetic crystals is given in [49, 50]. We shall mention only two general results of this theory, which distinguish the modern general scheme of the process (shown in Fig. 7) from the Van Vleck–de Krönig model. First, de Krönig and Van Vleck [2, 3] consider single-particle spin-lattice interaction, i.e., their calculations are carried out for a single isolated magnetic particle. Therefore, in their case the probability of a nonradiative transition is independent of the concentration of magnetic particles. Secondly, the experimentally measured relaxation time $\tau = \tau_{13}$ depends strongly on the temperature and this dependence is a very general phenomenon. Thus, for one-phonon spin-lattice transitions, we always have $\tau_{13} \approx T^{-1}$. On the other hand, the relaxation time τ for the mechanism involving the spin-

spin system (this time is measured in saturation experiments) may be completely independent of the lattice temperature; alternately, the temperature dependence of the relaxation time may be different from that given by the Van Vleck–de Krönig theory.

The theory of the relaxation mechanism of the establishment of a thermal equilibrium within the Zeeman system is developed in papers dealing with the spin-spin line width of the resonance absorption [38, 39]. The line width is governed by the terms $\hat{\mathcal{H}}^{exch}$ and $\hat{\mathcal{H}}_1^{dip}$ of the Hamiltonian (2.6), since only these two interactions can give rise to quantum transitions without a change in the Zeeman energy (the operator $\hat{\mathcal{H}}_2^{dip}$ has nonzero matrix elements only for transitions involving changes in the Zeeman energy and, therefore, it should be omitted in the theory of the spin-spin relaxation in the Zeeman system). When the inequality (2.7) is obeyed, $\hat{\mathcal{H}}^{dip}$ and $\hat{\mathcal{H}}_1^{dip}$ are small perturbations compared with $\hat{\mathcal{H}}^Z$. These interactions broaden the Zeeman levels and, consequently, the resonance absorption line. The contribution of these interactions to the line width governs the rate of the spin-spin relaxtion process.

The interactions $\hat{\mathcal{H}}^{exch}$ and $\hat{\mathcal{H}}_1^{dip}$ are not of equal importance in the theory of the spin-spin relaxation. The theory of the line width given in [38, 39, 41] shows that, if the magnetic dipole spin-spin interaction $\hat{\mathcal{H}}_1^{dip}$ broadens the absorption line, the isotropic exchange in the case $\hat{\mathcal{H}}^{exch} > \hat{\mathcal{H}}_1^{dip}$ reduces the dipole-dipole line width (the width for $\hat{\mathcal{H}}^{exch} = 0$) because of the averaging of the dipole spin-spin interaction of magnetic particles with their environment due to the exchange motion of the particles.

The exchange narrowing of the absorption line can be explained theoretically as follows. The exchange interaction does not contribute directly to the time dependence of the interaction operator (this dependence governs the absorption spectrum) because of commutation of the isotropic exchange operator with the operator representing interaction of the spins with the external radiation field. The time dependence of the interaction operator is solely due to the term $\hat{\mathcal{H}}_1^{dip}$. However, the exchange still affects the absorption spectrum since it gives rise to a time dependence $\hat{\mathcal{H}}_1^{dip}$ because $[\hat{\mathcal{H}}^{exch}, \hat{\mathcal{H}}_1^{dip}] \neq 0$. If $\hat{\mathcal{H}}^{exch} > \hat{\mathcal{H}}_1^{dip}$, the characteristic frequency of the exchange motion is higher than the characteristic frequency of the magnetic dipole-dipole interaction and the electron exchange results in an effective averaging of the dipole interaction of the spins with their environment and a consequent narrowing of the absorption line. The formula for the spin-spin line width is given in Chapter I [Eq. (1.2)]. Experimental data and theoretical estimates show that the relaxation times τ_1 and τ_{13} of various substances depend on temperature, concentration of magnetic particles, and other experimental conditions; the values of these relaxation times lie within the intervals:

$$\tau_1 \approx 10^{-7} - 10^{-10} \text{ sec},$$
$$\tau_{13} \approx 1 - 10^{-9} \text{ sec}.$$

We shall now consider the second limiting case of the theory of paramagnetic relaxation corresponding to the flow of energy through the spin-spin system.

The experimentally observed recovery of the equilibrium in the Zeeman system of levels is a function of four relaxation times: the times τ_1 and τ_2 for the establishment of equilibria within the Zeeman and exchange systems, the time τ_{12} for the establishment of equilibrium between the Zeeman and exchange spin-degrees of freedom, and the time τ_{23} characterizing the coupling of the exchange degrees of freedom with the thermal vibrations of the lattice. The relaxation process depends on relationships between these time constants. Two cases are of interest: a) $\tau_1 \ll \tau_{12}$ and b) $\tau_1 \approx \tau_{12}$ for an arbitrary value of the ratio τ_{12}/τ_{23}. In both cases, we shall assume that $\tau_2 \ll \tau_{12}, \tau_{23}$, i.e., the exchange system is always in thermal equilibrium characterized by a temperature $T_2 \geq T_0$.

If $\tau_1 < \tau_{12}$ an equilibrium within the Zeeman or exchange systems is established more rapidly than an equilibrium with other types of degrees of freedom. The state of the Zeeman system is then given by a temperature $T_1 \geq T_2 \geq T_0$. Under these conditions the continuous saturation experiments are described, by analogy with Eq. (2.9), by the following system of equations

$$P = \frac{E_1(T_1) - E_1(T_2)}{\tau_{12}} = \frac{E_2(T_2) - E_2(T_0)}{\tau_{23}}. \quad (2.11)$$

Here the notation is partly the same as that used in Eq. (2.9). The quantity $E_2(T_2)$ is the energy of the exchange system of degrees of freedom. By analogy with the Zeeman spin system, this energy can be written in the form

$$E_2(T_2) = \text{const} - \frac{A_2}{kT_2} \quad \text{for} \quad A_2 = \frac{1}{2} NzJ^2,$$

where J is the isotropic exchange interaction constant, z is the number of nearest neighbors of a given spin in the crystal lattice. Using the notation

$$X = \frac{T_0}{T_1}, \quad Y = \frac{T_0}{T_2},$$

$$\alpha_{12} = \frac{A_1}{\tau_{12}}, \quad \alpha_{23} = \frac{A_2}{\tau_{23}}, \quad s = 2W\tau_{12},$$

we can write Eq. (2.11) in the form

$$s\alpha_{12}X = \alpha_{12}(Y - X) = \alpha_{23}(1 - Y).$$

Here α_{12} and α_{23} are coefficients which represent the thermal contact between the corresponding systems. The solution of this system of equations gives:

$$X = \frac{1}{1 + s\left(1 + \frac{\alpha_{12}}{\alpha_{23}}\right)}, \quad Y = \frac{1 + s}{1 + s\left(1 + \frac{\alpha_{12}}{\alpha_{23}}\right)}. \quad (2.12)$$

Comparison of these results with the formula for the saturation factor (2.4) shows that the characteristic time τ for the establishment of an equilibrium between the Zeeman and other systems is

$$\tau = \tau_{12}\left(1 + \frac{\alpha_{12}}{\alpha_{23}}\right) = \tau_{12}\left(1 + \frac{A_1}{A_2}\frac{\tau_{23}}{\tau_{12}}\right). \quad (2.13)$$

If the thermal contact between the Zeeman and exchange systems is poorer than the thermal contact between the exchange system and the lattice ($\alpha_{12} \ll \alpha_{23}$), we find that

$$X = \frac{1}{1+s}, \quad Y = 1,$$

and the temperature of the exchange system $T_2 \simeq T_0$ is approximately equal to the lattice temperature. In this case the experimentally measured relaxation time τ is governed by the temperature-independent relaxation time τ_{12} for the interaction between the Zeeman and exchange spin systems

$$\tau \approx \tau_{12}.$$

It is interesting to note also that there is a temperature above which no increase in the amplitude of the hf field can heat the exchange system. When $s \to \infty$, it follows from Eq. (2.12) that

$$T_2 \to T_0\left(1 + \frac{\alpha_{12}}{\alpha_{23}}\right).$$

In the other case of interest to us, we have $\tau_1 \approx \tau_{12}$ and the macroscopic view of the paramagnetic relaxation is quite different. The Zeeman and exchange systems should now be considered thermodynamically as a single spin system in which an equilibrium is established in a time interval $\tau_1 \approx \tau_{12}$. As in the preceding case, the time τ determined from experiments on saturation (cf. § 1) depends on the ratio of two thermal contact coefficients: α_1 representing the internal thermal conductivity of the spin system, and α_{23}, representing the thermal contact between the spin-spin degrees of freedom and the lattice. When the ratio of these coefficients is small, $\alpha_1/\alpha_{23} \ll 1$, the relaxation time τ is governed by the temperature-independent time necessary for the establishment of an internal equilibrium in the Zeeman system: $\tau_1 \approx \tau_{12}$. This case applies to the free radical DPPH. The relaxation times found from saturation experiments and from measurements of the line width of DPPH are independent of temperature and approximately equal.

In conclusion, we shall consider briefly the results of the quantum theory of the relaxation times τ_1, τ_2, τ_{12}, and τ_{23} in order to determine the nature of the interactions responsible for nonradiative transitions in a crystal.

The time τ_{23} is associated with the spin-phonon interaction of the exchange system and the lattice. This interaction is due to a strong dependence of the exchange integral on the distance between particles. The formula for the probability of a one-phonon relaxation process is given in [34]:

$$\frac{1}{\tau_{23}} = \frac{3}{\pi}(z-1)^2 J^4 \frac{\lambda^2 a^2}{\rho v^5 \hbar^4} k T_0.$$

Here ρ is the density of the crystal; v is the velocity of sound; a is the distance between exchange-coupled spins; λ is a constant in the expression giving the dependence of the exchange integral on the distance between interacting particles: $J \approx \exp(-\lambda a)$.

The time constants τ_1, τ_2, and τ_{12} are governed by the spin-spin interactions $\hat{\mathcal{H}}^{\text{exch}}$ and $\hat{\mathcal{H}}^{\text{dip}} = \hat{\mathcal{H}}_1^{\text{dip}} + \hat{\mathcal{H}}_2^{\text{dip}}$ in the Hamiltonian of Eq. (2.6). The nature of the nonradiative spin-spin transitions characterized by the time constant τ_1 has already been considered in the first limiting case of the relaxation theory, when $\hat{\mathcal{H}}^z > \hat{\mathcal{H}}^{\text{exch}}, \hat{\mathcal{H}}^{\text{dip}}$ and the flow of energy through the spin-spin system can be neglected compared with the direct flow of energy from the Zeeman system to the lattice. In the other limiting case of the relaxation theory the flow of energy through the spin-spin system is important. This flow gives rise, generally speaking, to heating of the spin-spin reservoir, i.e., it increases the internal energy of the reservoir and alters its quantum state. There is a corresponding change in the contribution of the spin-spin interactions to the resonance line width or, in other words, there is a change in the rate of relaxation within the Zeeman system. The theory of this effect is complex. However, in the case of high spin-spin energy considered here [cf. inequality (2.8)]

$$\hat{\mathcal{H}}^{\text{exch}} + \hat{\mathcal{H}}_1^{\text{dip}} > \hat{\mathcal{H}}^z$$

and, because the specific heat of the spin-spin reservoir is large, an increase in the temperature of the reservoir is slight $[(T_2 - T_0) \ll T_0]$ compared with the Zeeman energy. Therefore, the influence of heating of the spin-spin system on the relaxation time τ_1 can be neglected.

Moreover, since the theory of the spin-spin resonance line width [38, 39] includes only the ratio of the energies $\hat{\mathcal{H}}^{exch}/\hat{\mathcal{H}}^{dip}$, but not the ratio $\hat{\mathcal{H}}^{exch}/\hat{\mathcal{H}}^{z}$, all the results are valid when the inequality (2.8) is obeyed.

According to [17, 39], the order of magnitude of τ_1 is given by the relationship [see also Eq. (1.2)]

$$\frac{h}{\pi \tau_1} \approx \frac{D^2}{h \nu_k}, \qquad (2.14)$$

where $D \approx g^2 \beta^2 / a^3$ is the constant of the magnetic dipole interaction of spins (a is the average distance between spins). The frequency ν_k represents the rate of averaging due to the exchange interaction of the effective magnetic field established at the site of a given spin by its magnetic dipole environment (the exchange narrowing effect). In the paramagnetic range of temperatures we have $h\nu_k \approx J$.

If the energy of the spin-spin system is comparable with the Zeeman energy (for example, if $\hat{\mathcal{H}}^{exch} = 0$, $\hat{\mathcal{H}}^z \gg \hat{\mathcal{H}}^{dip}$), the flow of energy through the spin system may heat it appreciably. This is manifested by a strong distortion of the line absorption profile. Effects of this kind are considered in [51].

Nonradiative transitions involving energy transfer from the Zeeman to the spin-spin degrees of freedom are characterized by a time constant τ_{12}. The two systems are coupled by the magnetic dipole interaction $\hat{\mathcal{H}}_2^{dip}$. The spectrum of the spin-spin system, consisting of the dipole-dipole and exchange degrees of freedom, is practically continuous because there is no correlation between the interactions of pairs of particles. The exchange energy is much higher than the dipole energy and, therefore, the dipole energy can be neglected in the calculation of τ_{12}. The probability of a transition (per unit time) involving a transfer of an energy quantum δ from the Zeeman to the exchange spin system under the action of $\hat{\mathcal{H}}_2^{dip}$ can be written in the same manner as the formulas for the probability of the cross-relaxation transition given by Eqs. (3)-(5) in [47]. The order of magnitude of this probability is

$$\frac{h}{\pi \tau_{12}} \approx \frac{D^2}{J} \exp\left(-m \frac{\delta^2}{J^2}\right), \qquad (2.15)$$

where $m = \text{const} \approx 1$. It follows from Eq. (2.15) that the probability $1/\tau_{12}$ has a maximum at $\delta/J \approx 1$ in its dependence on J. At high values of the exchange integral $J > \delta$, we have

$$\frac{h}{\pi \tau_{12}} \approx \frac{D^2}{J}$$

and comparison with Eq. (2.14) shows that the time constants τ_1 and τ_{12} are approximately equal. It is worth noting the analogy of these results with those obtained in the theory of nuclear paramagnetic relaxation in liquids [52].

The nature of the time τ_2 for the establishment of an equilibrium within the exchange system is self-evident. The time constant τ_2 is related to the exchange interactions and it can be calculated using formulas given in [47]. A correct estimate gives $1/\pi\tau_2 \approx (0.1-1)J$ and τ_2 is always smaller than τ_1 and τ_{12} (this has been assumed in the phenomenological theory).

Thus, the relaxation times τ_1, τ_2, and τ_{12}, characterizing the second relaxation mechanism, are governed by the spin-spin interactions. Consequently, they depend on the concentration of magnetic particles in a system and are independent or weakly dependent on the temperature. The time constant τ_{23} also depends on the exchange interaction and, therefore, it is a function of the concentration of magnetic particles. Thus, inclusion of the spin-spin degrees of freedom in the theory of relaxation gives rise to a concentration dependence of the experimentally determined relaxation time τ and to deviations from the universal temperature dependence of the rate of relaxation predicted by Van Vleck.

§ 3. Relaxation Through Spin-Spin Degrees of Freedom at Low Temperatures

Consideration of the measurements of the spin-lattice relaxation time of DPPH at liquid helium temperatures shows that it would be interesting to generalize the phenomenological theory of relaxation involving the spin-spin reservoir to the case of low temperatures, at which the specific heat of the lattice can no longer be regarded as infinite compared with the specific heat of the spin system. Experiments involving continuous saturation are described by the following standard system of equations:

$$P = \frac{E_1(T_1) - E_1(T_2)}{\tau_{12}} = \frac{E_2(T_2) - E_2(T_3)}{\tau_{23}} = \frac{E_3(T_3) - E_3(T_0)}{\tau_{34}}. \tag{2.16}$$

The subscripts 1, 2, 3, and 4 refer, respectively, to the Zeeman system, the exchange spin system, the lattice, and the thermostat (the container filled with liquid helium). Here $T_4 = T_0$ is the temperature of the thermostat. Assuming $\delta < kT_0$ and $J < kT_0$, we obtain (as in § 2) the following relationships for the energies of the subsystems

$$\left. \begin{array}{ll} E_1 = \text{const} - \dfrac{A_1}{kT_1} & \text{for} \quad A_1 = \dfrac{1}{4} N\delta^2, \\[4pt] E_2 = \text{const} - \dfrac{A_2}{kT_2} & \text{for} \quad A_2 = \dfrac{1}{2} NzJ^2, \\[4pt] E_3 = A_3(kT_3), & P = \dfrac{2WA_1}{kT_1}, \end{array} \right\} \tag{2.16a}$$

where A_3 is a constant independent of T_3. The rest of the notation is the same as in § 2. The system of equations (2.16) is valid if changes in the lattice temperature are small: $(T_3 - T_0 \lesssim \lesssim T_0$. In general, the expressions in Eq. (2.16) provide a very rough description of the relaxation process because: a) it is assumed that $\tau_1 < \tau_{12}$, although in fact $\tau_1 \approx \tau_{12}$ in the case of strong exchange; b) the energy of the exchange system is derived by analogy with the Zeeman energy which, generally speaking, is incorrect because the spectrum of the exchange system in the paramagnetic region is continuous; c) at high values of T_2 (§ 2) the value of τ_1 may depend on the power P but this has been ignored. In spite of these limitations, we can expect Eq. (2.16) to give us correct qualitative results. Using the notation

$$\begin{array}{lll} X = \dfrac{T_0}{T_1}, & Y = \dfrac{T_0}{T_2}, & Z = \dfrac{T_3}{T_0}, \\[6pt] \alpha_{12} = \dfrac{A_1}{\tau_{12}}, & \alpha_{23} = \dfrac{A_2}{\tau_{23}}, & \alpha_{34} = \dfrac{A_3}{\tau_{34}}, \end{array} \tag{2.17}$$

and

$$s = 2W\tau_{12},$$

we can write the system (2.16) in the form

$$s\alpha_{12}X = \alpha_{12}(Y - X) = \alpha_{23}Z\left(\frac{1}{Z} - Y\right) = \alpha_{34}(Z - 1). \tag{2.18}$$

Here α_{12}, α_{23}, and α_{34} are the thermal contact coefficients for the various subsystems taken at a temperature T_0. The derivation of Eq. (2.18) takes into account also the linear dependence of

α_{23} on T_3 [$\alpha_{23}(T_3) = Z\alpha_{23}(T_0)$] for one-phonon processes which establish an equilibrium between the exchange subsystem and the lattice at low temperatures (this will be discussed later). The quantities α_{12}, α_{23}, and α_{34} can depend also on the lattice temperature T_3 via the magnitude of the spin-spin interaction and the value of J (for example, the exchange integral of DPPH decreases when the temperature is lowered, as shown in Chapter I). However, this temperature dependence can be neglected partly because we are considering a small quantity $(T_3 - T_0)/T_0$.

The solution of the system (2.18) is of the form

$$Z = 1 + \frac{1 + s\left(1 + \frac{\alpha_{12}}{\alpha_{23}}\right)}{2(1+s)}\left[-1 + \sqrt{1 + \frac{4\frac{s\alpha_{12}}{\alpha_{34}}(1+s)}{\left[1 + s\left(1 + \frac{\alpha_{12}}{\alpha_{23}}\right)\right]^2}}\right], \quad (2.19)$$

$$X = \frac{\alpha_{34}}{s\alpha_{12}}(Z-1), \quad Y = \frac{\alpha_{34}(1+s)}{s\alpha_{12}}(Z-1).$$

The condition $(Z-1) < 1$ corresponds to the inequality

$$\frac{s\alpha_{12}}{2\alpha_{31}} < 1 + s\left(1 + \frac{\alpha_{12}}{2\alpha_{23}}\right). \quad (2.20)$$

Obviously, this condition is the main new result of our treatment. The stronger inequality

$$4\frac{s\alpha_{12}}{\alpha_{34}}(1+s) \ll \left[1 + s\left(1 + \frac{\alpha_{12}}{\alpha_{23}}\right)\right]^2$$

yields the well-known result for high temperatures discussed in § 2:

$$Z = 1, \quad X = \frac{1}{1 + s\left(1 + \frac{\alpha_{12}}{\alpha_{23}}\right)}, \quad Y = \frac{1+s}{1 + s\left(1 + \frac{\alpha_{12}}{\alpha_{23}}\right)}.$$

The saturation factor X [Eq. (2.19)], determined in continuous saturation experiments, can be expressed in terms of the quantities s, α_{12}, α_{23}, and α_{34}. We shall now consider the formula for the thermal contact coefficients.

It follows from the definition (2.17) that

$$\alpha_{12} = \frac{N\delta^2}{4\tau_{12}}. \quad (2.21)$$

The quantity $1/\tau_{12}$ can be expressed in terms of the line width $\Delta\nu$, whose values can be determined experimentally at various temperatures. The line width $\Delta\nu$ is given by the sum of the contributions

$$\Delta\nu = \frac{1}{\pi\tau_{12}} + \Delta\nu_0$$

representing the interactions involved in the establishment of an equilibrium between the Zeeman and exchange degrees of freedom as well as the spin-spin interactions which do not alter the Zeeman energy. It follows from Eqs. (2.14) and (2.15) that the line width is $\Delta\nu \approx 2/\pi\tau_{12}$ in the range $J > \delta$ of interest to us. A more rigorous calculation of the line width [17, 39, 40] shows that

$$\Delta\nu = \frac{10}{7}\left(\frac{1}{\pi\tau_{12}}\right) = \frac{10}{3}(\Delta\nu_0) \quad \text{for} \quad J \gg \delta. \quad (2.22)$$

Thus, the experimentally determined temperature dependence of the line width [21, 32] can be used to estimate α_{12} at various temperatures.

The coefficient α_{23} has been calculated in [34] for the one-phonon relaxation process. The second-order processes can be neglected at liquid-helium temperatures [34]:

$$\alpha_{23} = \frac{NzJ^2}{2\tau_{23}} = \frac{3}{2\pi} Z(Z-1)^2 J^6 \frac{N\lambda^2 a^2}{\rho v^5 \hbar^4} (kT_0). \tag{2.23}$$

The values of the time constants τ_{12} and τ_{23} of DPPH are discussed also in [53].

The transfer of energy from the exchange system to the lattice by the one-phonon relaxation process heats $\Delta Jn(J)$ lattice oscillators in the ΔJ band; the density of the energy distribution of the lattice oscillators is

$$n(J) = \frac{3VJ^2}{2\pi^2 \cdot \hbar^3 v^3}, \tag{2.24}$$

where V is the volume of a crystal. The coefficient α_{34}, which represents the rate of transfer of heat to the boundaries of a crystal, is governed by the two processes just discussed.

At low temperatures, when the nonlinearity of the lattice vibrations is so weak that elastic waves are propagated in a crystal without scattering, heat is transferred to the boundaries of a sample only by $\Delta Jn(J)$ elastic vibrations. The mean free path of elastic waves is of the order of the linear size of a crystal $l \approx \sqrt[3]{V}$ and

$$\alpha'_{34} = \frac{\Delta Jn(J)(kT_0)^2}{\tau'_{34}} \quad \text{where} \quad \tau'_{34} \approx \text{const} \frac{l}{v}. \tag{2.25}$$

The constant multiplier in the expression for τ'_{34} represents an increase in τ'_{34} because of the reflection of sound at the boundary of a crystal. We note that in this case the temperature T_3 refers to a small number $\Delta Jn(J)$ of excited oscillators and a change in this temperature does not alter the value of the exchange integral of DPPH.

At high temperatures we find that, due to the anharmonicity of the lattice vibrations, $\Delta Jn(J)$ excited oscillators reach equilibrium with all the lattice oscillators (which are then included in the heat conduction mechanism) in a short time given by [54]

$$\tau''_{34} \approx 1.5 \cdot 10^{-20} \frac{\rho v^5}{\gamma^2} \left(\frac{\hbar}{JT_0^4}\right) \tag{2.26}$$

(here $\gamma \approx 1$ is the Grüneisen coefficient). Assuming that in this case the rate of energy transfer is more likely to be limited by the process of the establishment of an equilibrium frequency distribution of the lattice oscillators than by the thermal conductivity of the lattice, we obtain

$$\alpha''_{34} = \frac{\Delta Jn(J)(kT_0)^2}{\tau''_{34}}. \tag{2.27}$$

Obviously, $\alpha_{34} = \alpha'_{34} + \alpha''_{34}$. These considerations will be used in the next section to discuss the experimental data on the paramagnetic relaxation of DPPH free radicals at liquid-helium temperatures.

TABLE 4

Orientation	Concentration	
	$2.1 \cdot 10^{-3}$	$4.6 \cdot 10^{-3}$
$H_0 \parallel a$	9.5±1.0	—
$H_0 \parallel b$	10.8±1.6	—
$H_0 \parallel c$	14.5±1.3	18±1.3

The spin-lattice relaxation time was calculated from the formula

$$\frac{1}{\tau} = \left(\frac{X}{1-X}\right)\frac{k\gamma H_1^2}{\Delta H}, \qquad (3.2)$$

where $\gamma = 2\pi g \beta/h$ is the gyromagnetic ratio, k = 0.62 ± 0.12 (k = 0.74 for the Gaussian absorption line profile and k = 0.50 for the Lorentzian profile). The square of the amplitude of the hf magnetic field of 42 MHz frequency was $H_1^2 = 5 \cdot 10^{-4}$ Oe. This value was determined experimentally from the amplitude of the hf voltage U_0 in the circuit containing the sample. ΔH and X are, respectively, the experimentally determined absorption line width and the saturation factor.

The experiments were carried out as follows. The temperature dependences of the amplitude and width of the absorption line were determined using various (Fig. 10) values of U_0 by pumping liquid-helium vapor in the interval T = 4.2–1.5°K and heating a previously cooled sample in the interval T = 0.08–1.50°K. The spectrometer circuit made it possible to determine the absolute intensity of the absorption by comparing the resonance signal with a reference signal provided by a variable resistor connected in parallel with the circuit containing the sample. The absorption intensity, relative to the reference signal, was found to be independent of the hf magnetic field in the absence of saturation. The reference signal was used also to check the linearity of the spectrometer. The absorption line was observed on the screen of an oscilloscope together with the reference signal, and both traces were photographed. The static magnetic field was modulated sinusoidally at a frequency of 0.5 Hz (amplitude 25–35 Oe) in order to observe the resonance absorption line on the oscilloscope screen. It was found that within the limits of the experimental error, the absorption line width was independent of the temperature and of the amplitude of the applied hf field (the measured values of ΔH are given in oersteds in Table 4). The line width was $\Delta H \approx 12$ Oe for $f = 10^{-3}$, $H \parallel c$. Therefore, at all temperatures, the saturation factor was calculated from $X = I/I_0$, where I and I_0 are the absorption line amplitudes at the center of the line measured relative to the amplitude of the reference signal: I is the amplitude in the presence of a saturating hf magnetic field H_1, I_0 is the amplitude in the presence of an hf magnetic field which is too weak to produce saturation. A considerable saturation of the absorption line was always observed at the lowest temperatures even at the lowest possible hf voltage in the circuit $U_0 \approx 0.3$ V. Therefore, the value of I_0 at the lowest temperatures was found by extrapolation, using the T^{-1} law, from the values of I_0 obtained at the higher temperatures, at which a reduction of the hf voltage eliminated the saturation effect. The values of τ obtained from Eq. (3.2) for various values of H_1 agreed satisfactorily at any given temperature (Fig. 10).

The relative error in the measurements is represented by the scatter of the experimental points in Fig. 10. The systematic error was estimated to be $\Delta\tau/\tau \lesssim \pm 0.6$. The accuracy of the measurements of the temperature was better than 10%.

A crystal to be investigated was placed in the circuit so that the angle between one of its axes and the static magnetic field H did not exceed ± 5°. During the experiments, the direction of the static field was perpendicular to that of the hf magnetic field and it could be rotated by an arbitrary angle in the *ac* or *bc* planes. The values of the g factors were not specially determined. However, the ratio of the resonance fields in cases H||*c* and H||*a* or H||*b* was approximately 2 in all the experiments, in agreement with the results reported in [64].

The spectrometer used to observe EPR absorption below 1°K is described in detail in Chapter IV.

We shall conclude this section by giving a summary of the results obtained. Figure 10 shows that two temperature regions can be distinguished. At relatively high temperatures (T = = 4.2-2.0°K) the experimental data for the concentrations $f = 0.21$ and 0.10% were obtained for H||*c*. The relaxation time $\tau(4.2°K) \approx 10^{-3}$ sec for $f = 0.21\%$ did not differ greatly from $\tau(4.2°K) \approx 5 \cdot 10^{-4}$ sec obtained in experiments at about 10^4 MHz. The temperature dependence $1/\tau \sim T^9$ was observed at T = 4.2-2.25°K for $f = 0.1\%$ and the corresponding temperature dependence for $f = 0.21\%$ at T = 4.2-3.4°K was $1/\tau \sim T^7$. The very strong temperature dependence of the relaxation time in this range of temperatures indicates that the energy is transferred from the spins to the lattice mainly by the Raman scattering of phonons on paramagnetic ions (two-phonon processes). The difference between the probabilities of nonradiative transitions in samples with $f = 0.1$ and 0.21% near T = 4.2°K and the probability of similar transitions at $\lambda \approx 3$ cm indicates the presence (in addition to the spin-phonon interaction via the spin-orbit coupling, predicted by the theory of Van Vleck and de Krönig [2, 3]) of an interaction of the spins with the thermal vibrations of the lattice, in accordance with the Waller mechanism [12], via the concentration-dependent spin-spin mechanism (the nature of this interaction will be discussed later).

The most interesting and the most detailed experimental data were obtained in the temperature range T < 2°K. We note, first, that the relaxation time is constant for H||*c* at T < 2°K but that this constancy is replaced by the linear dependence $1/\tau \sim T$ at the very lowest temperatures (Fig. 10). Secondly, an increase in the concentration of paramagnetic ions by a factor of about 2 does not alter the nature of the temperature dependence of $1/\tau$ in the case of H||*c*, but it increases very strongly (by a factor of up to 100) the probability of nonradiative transitions. The data obtained for $f = 0.1\%$ and H||*c* at T ≈ 2°K show that, provided $1/\tau$ does not increase during cooling, the value of $1/\tau$ at this concentration in the range T < 2°K decreases compared with $f = 0.21\%$ by a factor of not less than 100. Finally, a change in the orientation of a crystal with $f = 0.21\%$ relative to the external static magnetic field alters basically the nature of the temperature dependence of $1/\tau$. The temperature dependence of the relaxation time is nearly linear for H||*a* and H||*b*. This indicates that the relaxation mechanisms are different for H||*c* and H||*a*.

At the lowest temperature, the heating of the lattice is not observed at the frequencies of one-phonon relaxation processes since the results deduced from the experimental data and Eq. (2.2) are independent of the hf field amplitude.

§2. Two Mechanisms of Spin-Lattice Interaction with Ion Pair Participation

We shall now consider the possible influence of the exchange interactions of paramagnetic ions on the spin-lattice relaxation.

In magnetically dilute crystals, the exchange interaction energy is equal to zero except for a relatively small number of ion pairs n_i located at neighboring or closely spaced lattice sites. At a low concentration of paramagnetic ions, $f \ll 1$, we have

$$n_i = \frac{N}{2} f z_i. \tag{3.3}$$

Here N is the total number of paramagnetic ions in a crystal; z_i is the number of possible relative positions of ions in a pair (the energy spectrum of the pair is the same for each of those positions). We shall denote such pairs by the subscript i. We shall consider the exchange spin system described by the Hamiltonian

$$\hat{\mathcal{H}} = \frac{1}{2} Nf \sum_i z_i \hat{\mathcal{H}}_i^{exch},$$

where $\hat{\mathcal{H}}_i^{exch}$ is the Hamiltonian of one pair of the i-th type. If ion pairs have a short relaxation time, they may play an important role in the transfer of energy from paramagnetic ions to the lattice even when their concentration is low. This possibility was first pointed out by Van Vleck [5]. Two mechanisms of the spin-lattice relaxation with participation of ion pairs are considered theoretically in [6-8].

The first mechanism is similar to the relaxation through an exchange reservoir or "bath" [4] considered in Chapter II, except that the exchange reservoir is now formed by ion pairs. The relaxation time τ in this mechanism is given by [see Eq. (2.13)]

$$\tau = \tau_{12}\left(1 + \frac{A_1}{A_2}\frac{\tau_{23}}{\tau_{12}}\right), \tag{3.4}$$

where τ_{12} is the temperature-independent characteristic time for the establishment of an equilibrium between the Zeeman spin system whose Hamiltonian is given by Eq. (3.1) and the exchange spin system; τ_{23} is the time characterizing the rate of nonradiative transitions from the exchange system to the lattice;

$$A_1 \propto N\nu_z^2, \quad A_2 \propto n\nu_{exch}^2 \tag{3.4a}$$

are temperature-independent constants in the expressions for the energies in the Zeeman and exchange systems respectively: $E = const - A/kT$. The transfer of the excitation energy from the spins in a magnetic field to the ion pairs takes place by cross-relaxation. Therefore, this mechanism is active when the spectrum of an ion pair includes frequencies so close to the Zeeman frequency that the difference $h(\nu_{exch} - \nu_z)$ is not much larger than the energy of the dipole spin-spin interaction of an ion pair with isolated ions. Cross-relaxation processes of higher orders are possible also when any of the frequencies in the spectrum of an ion pair is equal to the Zeeman transition frequency [6].

In the second mechanism of the spin-lattice relaxation with the participation of ion pairs, the only limitation on the ratio of the characteristic Zeeman and exchange frequencies is imposed by the inequality $\nu_{exch} > \nu_z$. The energy transfer from the Zeeman system to the lattice is now due to the simultaneous quantum transitions of an isolated spin which emits a photon $h\nu_z$, or of an ion pair which emits or absorbs a quantum $h\nu_{exch}$ and generates or absorbs a phonon of energy $h(\nu_{exch} \pm \nu_z)$. These transitions are represented by the wavy arrows in the energy level scheme of a system comprising an ion pair and an isolated ion (Fig. 11). In the absence of the magnetic dipole interaction between the isolated ion and the ion pair, the spin-phonon interaction

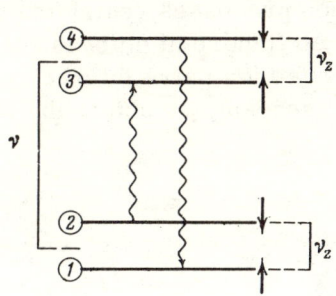

Fig. 11. Energy levels of a system consisting of an ion pair and an isolated ion in the $\nu_{exch} > \nu_z$ case. The vertical arrows indicate the state of an isolated ion.

of the ion pair cannot reverse the spin of the isolated ion. The forbiddenness of the spin reversal is lifted in the first approximation of the perturbation theory with respect to the parameter $(g^2\beta^2/r_0^3)/h\nu_z$. Consequently, the probability of a relaxation transition of an isolated spin is [6]

$$\frac{1}{\tau} \approx \frac{1}{\tau_{23}}\left(\frac{g^2\beta^2/r_0^3}{h\nu_z}\right)^2, \qquad (3.5)$$

where τ_{23} is the relaxation time of the ion pair. The possible large values of the ratio $\nu_{exch}/\nu_z > 1$ increase the rate of relaxation compared with the one-phonon relaxation at the frequency ν_z.

§3. Hypothesis on the Mechanism of Spin-Lattice Relaxation at Low Temperatures

The hypothesis which is now advanced to explain the experimental data runs as follows: one spin-lattice relaxation mechanism is active in the $H\|c$ orientation at concentrations $f = 0.21$ and 0.46%, and another mechanism applies to $H\|a$ and $H\|b$ at the iron ion concentration $f = 0.21\%$.

The linear temperature dependence of $1/\tau$ for the orientations $H\|a$, $H\|b$ reflects the temperature dependence $1/\tau$ governed by the one-phonon resonance energy transfer from the exchange pairs to the lattice at low temperatures. Because of the dependence $1/\tau$ on g^4 for $h\nu_z = $ const, this mechanism is much less effective in the $H\|c$ case, since $g_{cc} \approx 1$ and $g_{aa} = 2.29$, $g_{bb} = 2.16$. When the experimental data (Fig. 10) for the $H\|a$ case and Eq. (3.5) are used to calculate the g factor for the $H\|c$ orientation, we obtain values for $1/\tau$ which are about a fifth those observed experimentally for $H\|c$.

In the $H\|c$ case, the horizontal part of the dependence of $1/\tau$ on T corresponds to

$$\frac{A_1}{A_2}\frac{\tau_{23}}{\tau_{12}} < 1 \text{ and } \tau \approx \tau_{12}. \qquad (3.6)$$

At lower temperatures, at which the relaxation time increases approximately linearly with temperature, we have

$$\frac{A_1}{A_2}\frac{\tau_{23}}{\tau_{12}} > 1 \text{ and } \tau = \frac{A_1}{A_2}\tau_{23} = \frac{N}{n}\left(\frac{\nu_z}{\nu_{exch}}\right)^2 \tau_{23}. \qquad (3.7)$$

As before, the linear dependence of $1/\tau_{23}$ on temperature indicates resonance energy transfer at the frequency ν_{exch} (a one-phonon process), from the exchange system to the lattice.

This hypothesis allows us to explain qualitatively the strong concentration dependence of the relaxation time in the $H\|c$ orientation. At low paramagnetic ion concentrations, the cross-relaxation between the Zeeman and exchange transitions with appreciably differing frequencies is difficult and, therefore, we may find that at low temperatures the thermal contact between the exchange system and the lattice takes place by one-phonon processes restricted to only some of the frequencies in the exchange system spectrum. An important point to note is the absence of cross-relaxation within the exchange system because ion pairs interact weakly with one another. Using the frequency dependence $1/\tau_{23} \sim \nu_{exch}^2$, (see §5), we obtain the following expression for $1/\tau$ from Eq. (3.7):

$$\frac{1}{\tau} \approx f \frac{\nu_{exch}^4}{\nu_z^2}(kT). \qquad (3.8)$$

To explain the experimental data for the crystal with $f = 0.46\%$, we shall assume that the energy is transferred to the lattice mainly by quanta of energy $h\nu_{exch}''$ greater than $h\nu_{exch}'$ for the crystal with $f = 0.21\%$. The ratio $\nu_{exch}''/\nu_{exch}' = 2.2$ explains, in accordance with Eq. (3.8), the reduction in the relaxation time τ by a factor of 50. The experimentally determined relaxation times for these two concentrations differ approximately by a factor of 80 at $T = 0.2°K$.

This concentration dependence of the energy transfer from ion pairs to the lattice is possible in the case of a large reduction in the time constant τ_{12} of the cross-relaxation between the Zeeman transitions and the exchange transitions of frequency ν_{exch}'', compared with the time constant of the cross-relaxation between the Zeeman transitions and the exchange transitions of lower frequency ν_{exch}', when the concentration is increased from $f = 0.21\%$ to $f = 0.46\%$. According to the experimental determinations the time τ_{12} (Fig. 10) is reduced by a factor of approximately 50. Such a strong concentration dependence is typical of high-order cross-relaxation processes [6]. The relatively high absolute value of $1/\tau_{12}$ is reached in such cases only when the frequency of the exchange subsystem is a multiple of the Zeeman frequency. The spin diffusion can also contribute to the power dependence of τ_{12} on the concentration because this diffusion is responsible for the transfer of the excitation to the direct vicinity of an ion pair and it precedes the cross-relaxation transition, which transfers the excitation energy from an isolated ion to an ion pair.

If a sample is cooled so that its temperature is lower than the equivalent separation between the energy levels of an ion pair (participating in relaxation transitions) and the ground state, i.e., $kT < J$, we find that the number of pairs and, consequently, the probability of a relaxation transition in accordance with Eq. (3.7) both decrease exponentially when the temperature is lowered:

$$\frac{1}{\tau} \propto kT \exp\left(-\frac{J}{kT}\right). \tag{3.9}$$

This effect explains the superlinear rise of $1/\tau$ during cooling (Fig. 10) in the $H \| c$ case.

Thus, the hypothesis considered in the present section accounts qualitatively for the experimental observations at $T < 2°K$. To prove this hypothesis, we must determine the energy spectrum of the exchange system in potassium cobalticyanide, calculate theoretically the probabilities of the relaxation transitions, $1/\tau_{12}$ and $1/\tau_{23}$, and compare them with the experimental values. This will be done partly in the following sections.

§4. Energy Spectrum of Ion Pairs in Potassium Cobalticyanide

The exchange interactions of Fe^{3+} ions in cyanides are described in [64, 67, 68]. Measurements of the static magnetic susceptibility of $K_3Fe(CN)_6$ in the temperature range $T = 1.0$–$20°K$ are reported in [67]. Direct observations of the EPR spectrum of ion pairs in $K_3(Fe,Co) \cdot (CN)_6$ with a concentration of iron $f \approx 5\%$ are reported in [68]. The lattice parameters of these two cyanides are practically equal [69]. The experiments can be interpreted by means of the Hamiltonian

$$\hat{\mathcal{H}}_{12} = J\hat{s}_1\hat{s}_2 + \hat{s}_1\Gamma\hat{s}_2 - \beta Hg(\hat{s}_1 + \hat{s}_2). \tag{3.10}$$

The first term corresponds to an isotropic exchange interaction of ions with $s_1 = s_2 = \frac{1}{2}$; the second represents a symmetric anisotropic exchange interaction. The following results are obtained: $J = 0.29$ cm^{-1}; the principal axes of the tensor Γ correspond to the crystal axes a, b, and c; $\Gamma_a = -0.043$ cm^{-1}, $\Gamma_b = -0.021$ cm^{-1}, $\Gamma_c = +0.064$ cm^{-1}; $\Gamma_a + \Gamma_b + \Gamma_c = 0$. These values correspond to the nearest-neighbors positions of iron ions in the crystal lattice. It is found that the maximum number of neighbors (z_1) with which an ion can be coupled by the exchange interaction (of the value just given) is $z_1 = 2$. This fact, taken in conjunction with the crystal lattice structure, indicates that ion pairs with the maximum exchange interaction consist of Fe^{3+} ions located along the a axis of the unit cell [69]. The tetrahedral complexes $[Fe(CN)_6]^{3-}$ along the two other crystal axes are separated by K^+ potassium ions whose ionic binding with cyanide complexes excludes the possibility of electron exchange between iron ions along these axes. The maximum exchange interaction cannot be attributed either to ions at two nonequivalent positions within one unit cell since in this case we would have $z_1 = 4$. Other absorption lines are reported in [68]: they are evidently due to pairs of nearest nonequivalent ions or pairs of less closely spaced iron ions.

We shall analyze the experimental data using a spin Hamiltonian which is more general than that in Eq. (3.10):

$$\left.\begin{aligned}
\hat{\mathcal{H}}^S &= \hat{\mathcal{H}}^J + \hat{\mathcal{H}}^D + \hat{\mathcal{H}}^\Gamma + \hat{\mathcal{H}}^Z, \\
\hat{\mathcal{H}}^J &= \tfrac{1}{2} J \left[(S(S+1) - \tfrac{3}{2}) \right], \qquad \hat{\mathcal{H}}^D = \mathbf{D}\,[\mathbf{s}_1 \times \mathbf{s}_2], \\
\hat{\mathcal{H}}^\Gamma &= \tfrac{1}{2}(\Gamma_a \hat{S}_a^2 + \Gamma_b \hat{S}_b^2 + \Gamma_c \hat{S}_c^2), \\
\hat{\mathcal{H}}^Z &= -\beta \mathbf{H} g \mathbf{S}.
\end{aligned}\right\} \quad (3.11)$$

The difference between this new Hamiltonian and Eq. (3.10) lies in the addition of an antisymmetric anisotropic exchange coupling $\hat{\mathcal{H}}_D$ suggested in [70] and calculated quantum-mechanically in [71]. The total spin of a pair $\mathbf{S} = \mathbf{s}_1 + \mathbf{s}_2$ ($S = 0, 1$) is used in Eq. (3.11).

We shall now find the eigenvalues and eigenfunctions of the Hamiltonian (3.11) using the perturbation method and assuming that $\hat{\mathcal{H}}^J \gg \hat{\mathcal{H}}^D + \hat{\mathcal{H}}^\Gamma + \hat{\mathcal{H}}^Z$. In the zeroth approximation of the theory, we have two energy levels: a lower level $E_0 = -3J/4$, corresponding to the spin $S = 0$, and a triply degenerate upper level $E_1 = J/4$ with $S = 1$. The eigenfunctions of the zeroth approximation will be denoted by $|SM\rangle$, where M is the projection of the vector $\mathbf{S}$ on the quantization axis. A characteristic of the perturbation theory used in this calculation is the equality $\langle 1, M' | \hat{\mathcal{H}}^D | 1, M \rangle = 0$ for any values M, M' = 0 ± 1. Since $\hat{\mathcal{H}}^D > \hat{\mathcal{H}}^\Gamma, \hat{\mathcal{H}}^Z$ [71], the perturbation theory must include the terms which are quadratic in $\hat{\mathcal{H}}^D$ as well as the terms linear in $\hat{\mathcal{H}}^\Gamma$, and $\hat{\mathcal{H}}^Z$ [72, 73]. For the sake of simplicity, we shall assume that

$$D_c = D, \quad D_a = D_b = 0. \tag{3.12}$$

Some justification for this assumption is provided by the theory given in [71], according to which $D_c \sim |2 - g_{cc}|$ is larger than $D_a \sim |2 - g_{aa}|$, because $g_{cc} = 0.91$ and $g_{aa} = 2.29$. Consequently, we may assume that the energy spectrum of an ion pair is not greatly affected if D_a, $D_b \neq 0$. The splitting of the triplet E_1 by a perturbation in Eq. (3.11) in the cases $\mathbf{H} \parallel a$ and $\mathbf{H} \parallel c$, corresponding to the experimental conditions, is given by the following formulas [72, 73], which are accurate to terms of the second order:

$$\mathbf{H} \parallel a, \quad \mathcal{E}_{\mathrm{I,III}} = \left(\frac{\Gamma_a}{4} + \frac{D^2}{8J}\right) \pm \sqrt{\delta^2 + \left(\frac{\Gamma_c - \Gamma_b}{4} - \frac{D^2}{8J}\right)^2}, \quad \mathcal{E}_{\mathrm{II}} = -\frac{\Gamma_a}{2}, \quad \delta = \beta g_{aa} H, \tag{3.13'}$$

$$\mathbf{H}_0 \parallel c, \quad \mathcal{E}_{\mathrm{I,III}} = \frac{\Gamma_c}{4} \mp \sqrt{\delta^2 + \frac{(\Gamma_b - \Gamma_a)^2}{16}}, \quad \mathcal{E}_{\mathrm{II}} = -\frac{\Gamma_c}{2} + \frac{D^2}{4J}, \quad \delta = \beta g_{cc} H. \tag{3.13''}$$

The values of the energies $\mathcal{E}_I$, $\mathcal{E}_{II}$, $\mathcal{E}_{III}$ are measured from $E_1 = J/4$. The corresponding eigenfunctions of the states with $S = 1$ are given by the following formula, which is accurate to terms of the first order:

$$|K\rangle = \sum_M C_{KM}\left[|1, M\rangle + \frac{\langle 0, 0|\hat{\mathcal{H}}^D|1, M\rangle}{J}|0, 0\rangle\right] \quad (K = \text{I, II, III}). \tag{3.14}$$

The coefficients C_{KM} are defined by the following relationships for the zeroth order functions (the quantization axis is assumed to lie along the static magnetic field):

$$\text{I}) = N|1, 1\rangle + M|1, -1\rangle, \quad \text{II}) = |1, 0\rangle, \quad \text{III}) = -M|1, 1\rangle + N|1, -1\rangle. \tag{3.15}$$

Assuming that $\Gamma_c > \Gamma_b > \Gamma_a$ [68] and $(\Gamma_c - \Gamma_b)/4 > D^2/8J$, we obtain for $\mathbf{H}\|a$

$$M = \left[1 + \frac{\left(\frac{\Gamma_c - \Gamma_b}{4} - \frac{D^2}{8J}\right)^2}{\left(-\delta + \sqrt{\delta^2 + \left(\frac{\Gamma_c - \Gamma_b}{4} - \frac{D^2}{8J}\right)^2}\right)^2}\right]^{-\frac{1}{2}},$$

$$N = \left[1 + \frac{\left(-\delta + \sqrt{\delta^2 + \left(\frac{\Gamma_c - \Gamma_b}{4} - \frac{D^2}{8J}\right)^2}\right)^2}{\left(\frac{\Gamma_c - \Gamma_b}{4} - \frac{D^2}{8J}\right)^2}\right]^{-\frac{1}{2}}; \tag{3.16'}$$

for $\mathbf{H}\|c$, we have

$$M = \left[1 + \frac{\frac{1}{16}(\Gamma_b - \Gamma_a)^2}{\left(-\delta + \sqrt{\delta^2 + \frac{1}{16}(\Gamma_b - \Gamma_a)^2}\right)^2}\right]^{-\frac{1}{2}},$$

$$N = \left[1 + \frac{\left(-\delta + \sqrt{\delta^2 + \frac{1}{16}(\Gamma_b - \Gamma_a)^2}\right)^2}{\frac{1}{16}(\Gamma_b - \Gamma_a)^2}\right]^{-\frac{1}{2}}. \tag{3.16''}$$

We note that $N = M = 1/\sqrt{2}$ for $\delta = 0$ and $M \to 0$, $N \to 1$ for $\delta \to \infty$.

The energy levels of an ion pair corresponding to the spin $S = 1$ ($E_1 = J/4$) are given in Fig. 12 for two orientations of the static magnetic field relative to the crystal.

We shall now consider nonradiative transitions in which only the levels shown in Fig. 12 participate. It follows from the experimental data that the triplet-singlet transitions make no contribution to the measured spin-lattice relaxation of iron ions. The linear temperature dependence of the rate of the one-phonon relaxation process observed down to $T = 0.1°$K (Fig. 10)

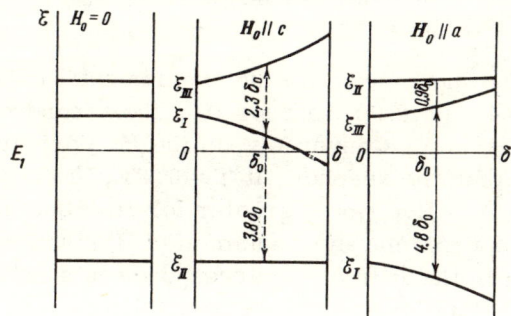

Fig. 12. Energy level scheme of an ion pair in the S = 1 state, shown as a function of the Zeeman splitting of an isolated ion δ. The arrows indicate the transitions playing the dominant role in the energy transfer from isolated ions to the lattice. In the experiments carried out by the present author, $\delta_0 = 0.002°$K.

indicates [see Eqs. (3.4)-(3.7)] that nonradiative transitions in the spectrum of a pair take place at frequencies corresponding to kT $\gg$ h$\nu_{exch} \approx$ 0.01-0.03, which are considerably lower than the frequencies of transitions between the states S = 1 and S = 0 of the ion pairs considered here. The slight contribution of the nonradiative singlet-triplet transitions to the first relaxation mechanism (§ 2) is due to the large (a factor of 10-100) difference between the frequencies of the Zeeman and exchange transitions. The slight contribution of the nonradiative transitions to the second mechanism is evidently due to the small value of the constant D in the Hamiltonian (3.11).

The nonradiative transitions between the triplet levels of ion pairs with the strongest exchange interactions (J = 0.42°K, Γ_a = −0.06°K, Γ_b = −0.03°K, Γ_c = 0.09°K, D $\lesssim$ 0.14°K) also fail to explain the experimental data. The upper limit of the value of D, which has just been given, does not contradict the results reported in [68]. According to the mechanism described in § 2, the pairs with the strongest exchange interaction are not observed experimentally for the following reasons. First, the Zeeman splitting of the levels of an isolated ion, δ_0 = 0.002°K, is much less than the minimum splitting of the levels of a pair in H = 0, which is 0.015°K. Therefore, in order to transfer energy from the Zeeman to the exchange system, we must postulate a cross-relaxation process of at least the seventh or eighth order, which is not compatible with the relatively small absolute value of the time constant τ_{12}, found experimentally. Secondly, the exponential rise of the relaxation time, expected on the basis of Eq. (3.9) is not observed for H$||$c even at kT $\approx$ J = 0.42°K (Fig. 10). A rapid rise of τ, due to a reduction in the population of levels with spin S = 1, is not observed for H$||$a at kT $\approx$ J = 0.42°K.

We shall now consider ion pairs with a weaker exchange coupling, for example, pairs of ions separated along an axis by a distance equal to two lattice constants. The value J $\approx$ 0.06°K for such pairs, predicted by Eq. (3.9), is not in conflict with the experimental dependence τ(T°K) for f = 0.21% in the H$||$c orientation at the lowest temperatures. However, since the scatter of the experimental points in Fig. 10 is considerable, this value of J should be regarded only as an estimate. The constants D and Γ are proportional to J [71] and, therefore, we shall assume that for the pairs of the type considered here, the respective values of these two constants are also approximately seven times smaller than those for the ion pairs with the strongest possible exchange coupling. The spectrum of a pair with a weaker coupling and with a Zeeman splitting δ_0 = 0.002°K, which was constant in our experiments, depends strongly (Fig. 12) on the direction of the static magnetic field relative to the crystal axes. Estimates of the separations between the levels are given in Fig. 12.

The energy level scheme in Fig. 12 makes it possible to consider in detail the spin-lattice relaxation mechanism and to find transitions in the exchange system spectrum corresponding to the energy transfer from ion pairs to the lattice. Thus, in the H$||$c case, the one-phonon energy transfer in a crystal with f = 0.21% can take place at the lowest frequency of the exchange system (the frequency of the transition $\mathcal{E}_{III} - \mathcal{E}_I = 2.3\delta_0$) and the corresponding energy transfer in a crystal with f = 0.46% can take place also at the frequency $\mathcal{E}_I - \mathcal{E}_{II} = 3.8\delta_0$ [the first relaxation mechanism given by Eq. (3.4)]. In the H$||$a case, the second relaxation mechanism [Eq. (3.5)] in a crystal with f = 0.21% is manifested by transitions at the highest possible frequency: $\mathcal{E}_{III} - \mathcal{E}_I = 4.8\delta_0$ or $\mathcal{E}_{II} - \mathcal{E}_I = 5.7\delta_0$.

This conclusion must be checked by a special comparison with the experimental data. We must check, using the curves in Fig. 10 and Eqs. (3.4) and (3.5), that for H$||$c the second relaxation mechanism is less effective than the first mechanism. Conversely, in the H$||$a orientation for f = 0.21%, the first mechanism is much weaker than the second. At T = 0.2°K, the probability of a nonradiative transition for f = 0.21% (Fig. 10) is five times greater for H$||$a than for H$||$c. Since the maximum splitting of the triplet is approximately the same for H$||$c and H$||$a, it follows from Eq. (3.5) that the second mechanism in the H$||$c case corresponds to a relaxation transition probability which is a factor of $(g_{aa}/g_{cc})^4 \approx$ 25 lower than the probability in the H$||$a case, i.e., at T = 0.2°K, the second mechanism in the $\mathbf{H}\|c$ case is at least five times less effective than the first mechanism. On the other hand, the first relaxation mechanism for the $\mathbf{H}\|a$ case may involve transitions at the frequency $\mathscr{E}_{II} - \mathscr{E}_{III} = 0.9\delta_0$. However, the corresponding probability of a nonradiative transition is, according to Eq. (3.8), lower than the probability of a transition in the $\mathbf{H}\|c$ orientation, i.e., it is unimportant compared with the rate of energy transfer due to the second mechanism in the $\mathbf{H}\|a$ orientation. Thus, the energy spectrum of the exchange system in the cyanide satisfies qualitatively the spin-lattice relaxation mechanism proposed in § 3.

We must bear in mind the possibility of some deviation from the simple relaxation process based on the Hamiltonian of Eq. (3.11). This is due to the possibility of a more complex spectrum of the exchange system in the cyanide. Such a possibility is supported by the absorption lines of ion pairs, which have not been interpreted in [68].

§ 5. Calculation of the One-Phonon Spin-Lattice Interaction of Exchange-Coupled Iron Ions. Quantitative Comparison of the Theory with Experiment

We shall estimate the probabilities of nonradiative transitions for an ion pair. We shall consider only the one-phonon spin-lattice relaxation which gives rise to a linear dependence of the relaxation probability on the temperature and which predominates at low temperatures. Lattice vibrations influence the spin in two ways: through the spin-spin interaction (the Waller mechanism), which depends on the relative positions of ions in a pair, and through the spin-orbit coupling (the Van Vleck–de Krönig) mechanism.

We shall consider first the Waller mechanism. If we assume that $D = \Gamma \approx J$ [71] and that $J \propto \exp(-\xi R_0)$, where R_0 is the distance between the ions in a pair, we find that the one-phonon spin-lattice interaction operator is

$$\hat{V} = (\hat{\mathscr{H}}^J + \hat{\mathscr{H}}^D + \hat{\mathscr{H}}^\Gamma)\,\xi\delta\hat{R}_0 = \hat{\tilde{V}}\xi(\delta\hat{R}_0). \tag{3.17}$$

Expanding the relative displacement of ions in a pair, δR_0, along the normal vibrations of the lattice and using certain well-known relationships [49] as well as the $h\nu_{exch} \ll kT$ approximation, we find that the probability of a nonradiative transition in the spectrum of an ion pair is

$$\frac{1}{\tau_{23}} \approx \frac{96\pi^3 \nu_{exch}^2 kT}{h^2 c^5 \rho}(\xi R_0)^2 |\langle K'|\tilde{V}|k\rangle|^2, \tag{3.18}$$

where ν_{exch} is the frequency of the one-phonon process; c is the velocity of sound; ρ is the macroscopic density of the crystal. According to Eq. (3.14), the matrix element is

$$\langle K'|\tilde{V}|K\rangle = \sum_{M'M} C^*_{K'M'}C_{KM}[\langle 1, M'|\mathscr{H}^\Gamma|1, M\rangle + \frac{2}{J}\langle 1, M'|\mathscr{H}^D|0, 0\rangle\langle 0, 0|\mathscr{H}^D|1, M\rangle]. \tag{3.19}$$

Using the assumptions represented by Eq. (3.12), we find that Eqs. (3.15) and (3.16) yield the only nonzero matrix element

$$\mathbf{H}\|a,\ \langle I|\tilde{V}|III\rangle = -(N^2 - M^2)\left(\frac{\Gamma_c - \Gamma_b}{4} - \frac{D^2}{4J}\right),$$

$$\mathbf{H}\|c,\ \langle I|\tilde{V}|III\rangle = -(N^2 - M^2)\left(\frac{\Gamma_b - \Gamma_a}{4}\right) \tag{3.20}$$

(this element is equal to zero when $\delta = 0$ because then $N = M$). The forbiddenness can be lifted from other transitions by assuming that $D_a, D_b \neq 0$. This should not affect the order of magnitude of the quantity in Eq. (3.20). The application of Eqs. (3.18)–(3.20) to $\mathbf{H} \| c$, for $T = 0.1°K$, $\nu_{\text{exch}} = 2.3\delta_0 = 10^8$ Hz, $c = 10^5$ cm/sec, $\rho = 1.85$ g/cm^3, $R_0 = 1.4 \cdot 10^{-7}$ cm, $\xi = 2.5 \cdot 10^7$, $\delta_0 = 0.002°K$ yields the value

$$\frac{1}{\tau_{23}} \approx 1.5 \cdot 10^{-7} \text{ sec}^{-1}.$$

This low probability disagrees strongly with the experimental data.

We shall now estimate the order of magnitude of the probability of a one-phonon relaxation transition in the spectrum of an ion pair for the case of the Van Vleck–de Krönig spin-lattice interaction. The problem is formulated as in the case of an isolated ion considered in [49]. The operators of the one-particle electron-phonon interaction, $\hat{V}_1$ and $\hat{V}_2$, are (in the one-phonon relaxation) the terms of an expansion of the energy of ions in a pair subjected to a crystal field; these terms are linear with respect to the normal displacements of the nearest neighbors. The electron-phonon interaction operator of an ion pair $\hat{V} = \hat{V}_1 + \hat{V}_2$ is independent of the spin variables. Therefore, admixtures of the excited states of ions of a pair in a crystal field, which appear due to the spin-orbit interaction, are important in the spin-lattice relaxation between the spin levels A and A' of the ground orbital state of a pair.

Since, at low frequencies, the exchange interaction of ions in a pair is stronger than the interaction of ions with an external static magnetic field, we can regard an ion pair in the theory of [49] as a particle with a spin $s = 1$. In this case, the matrix element of the electron-phonon interaction operator has the following order of magnitude for a transition between the spin levels of the ground state of a pair [Eq. (42) in [49]]:

$$\langle A' | V | A \rangle \approx \frac{4\lambda^2}{\Delta^2} V_1, \qquad (3.21)$$

where λ is the spin-orbit interaction constant and Δ is the difference between the energies of the ground and first excited states of an ion pair in the crystal field.

The value of the matrix element for an ion pair is considerably larger than the corresponding matrix element for an isolated iron ion with spin $s = \frac{1}{2}$, whose order of magnitude is

$$\langle A' | V_1 | A \rangle \approx \frac{\lambda (g\beta H)}{\Delta^2} V_1.$$

Here $g\beta H$ is the Zeeman splitting of the energy levels of an isolated ion. We shall show later that a difference by a factor of $\lambda / g\beta H$ is sufficient to explain the experimental data.

Using Eq. (3.21) and the formulas from the theory of the spin-lattice relaxation of an isolated iron ion [49], we can express the relaxation time of an ion pair τ_{23} in terms of the relaxation time of an isolated ion τ^*:

$$\frac{1}{\tau_{23}} \approx \left(\frac{1}{\tau^*} \right) 16 \frac{\nu_{\text{exch}}^2 \lambda^2}{\nu_z^4}.$$

The experimental results [61] indicate a one-phonon spin-lattice Van Vleck–de Krönig interaction of an isolated iron ion with $(1/\tau^*) = 5.4\,T$ at the Zeeman frequency $\nu_z = 8.7 \cdot 10^9$ Hz. Consequently, for $\mathbf{H} \| a$, when $\nu_{\text{exch}} = 4.8\delta_0 = 2 \cdot 10^8$ Hz and $\lambda = 280$ cm^{-1} [61], we find that

$$\frac{1}{\tau_{23}} = 4 \cdot 10^4 \, T,$$

which corresponds to $1/\tau_{23} = 4 \cdot 10^3$ sec^{-1} at $T = 0.1°K$. In the $\mathbf{H} \| c$ case, we have $\nu_{\text{exch}} = 2.3\delta_0$ and, consequently, $1/\tau_{23} = 10^3$ sec^{-1} at $T = 0.1°K$. In the latter case, the measured relaxation time of isolated ions is related to τ_{23} by the following expression ($f = 2.1 \cdot 10^{-3}$, $\nu_{\text{exch}} = 2.3\nu_z$), which is based on Eq. (3.7):

$$\frac{1}{\tau} = \frac{1}{\tau_{23}} \left(\frac{\nu_{exch}}{\nu_z}\right)^2 = 11 \text{ sec}^{-1}.$$

This result is in agreement with the experimental value $1/\tau = 8$ sec^{-1}, obtained under these conditions (Fig. 10). It also follows from Eq. (3.5) that for H$||a$ and T = 0.1°K the value $1/\tau_{23} = 4 \cdot 10^3$ sec^{-1} does not contradict the experimental value $1/\tau = 60$-70 sec^{-1} obtained under these conditions.

Thus, allowance for the electron-phonon interaction of an ion pair in accordance with the Van Vleck–de Krönig mechanism permits us to calculate theoretically the probabilities of relaxation transitions, which are in satisfactory agreement with the experiment. This agreement confirms the important role of exchange-coupled iron ions in the transfer of energy from isolated ions to the thermal vibrations of the lattice.

Our theoretical analysis has not included the calculation of the cross-relaxation time τ_{12} in Eq. (3.4). To carry out this calculation, we must know more accurately the spectrum and the line widths corresponding to the transitions considered in the spectrum of ion pairs. Nevertheless, a comparison of the experimental data with the theory gives convincing support to the paramagnetic relaxation mechanism suggested in § 3 for potassium cobalticyanide at low temperatures.

We have mentioned in the Introduction that some investigators have studied the influence of exchange interactions on the spin-lattice relaxation in ruby. Temperature-independent spin-lattice relaxation in ruby is reported in [74]. Investigations of the spectrum and relaxation of exchange-coupled pairs of chromium ions in ruby are reported in [9, 10]. A theory of the influence of an exchange interaction of ions (which is weak compared with the Zeeman interaction) on the relaxation of isolated chromium ions in ruby has been developed in [11]. However, since ruby has a more complex energy spectrum, it is less convenient than potassium cobalticyanide for investigations of the mechanism of paramagnetic relaxation with the participation of exchange-coupled paramagnetic ions.

During the same period as the present author's studies, Rannestad and Wagner [75] investigated, at helium temperatures, the spin-lattice relaxation in K_3(Fe,Co)(CN)$_6$ at 8500 and 1800 MHz, using samples with iron concentrations from $f = 0.24$ to $f = 1.7\%$. Their measurements at 1800 MHz indicated a pronounced rise in the rate of relaxation with increasing concentration. The rate of relaxation ceased to rise at $f = 1.7\%$, at which concentration the rate was approximately equal with the values obtained from the measurements at 8500 MHz (at this frequency the rate of relaxation depended weakly on the concentration of iron ions). At high concentrations, the experimental curves representing the re-establishment of thermal equilibrium in a substance after the end of excitation were not simple exponential functions. A possible explanation of these results, suggested by Rannestad and Wagner, is based on the assumption that, at high concentrations of iron ions ($f \gtrsim 1\%$) the excitation energy of the Zeeman levels is transferred by the cross-relaxation mechanism to exchange-coupled iron ion pairs, and from these pairs to the lattice. In concentrated samples, this relaxation mechanism may compete with the Van Vleck–de Krönig mechanism. The large number of pairs with different exchange interactions found in samples with high concentrations makes the spectrum of the exchange subsystem extremely complicated. The spin-lattice relaxation in such a system may be independent of the Zeeman frequency within a wide range of frequencies. Rannestad and Wagner [75] regard the concentrated solution of iron in potassium cobalticyanide ($f \approx 1\%$) as an extremely complex and inconvenient subject for investigating the influence of exchange interactions on the spin-lattice relaxation.

Recently, V. A. Atsarkin (Institute of Radio Engineering and Electronics, Academy of Sciences of the USSR) investigated, at 10,000 MHz, the low-temperature paramagnetic spin-lattice relaxation of Cr^{3+} ions in magnesium tungstate ($MgWO_4$) with chromium concentrations of about 10^{-3}. The temperature-independent spin-lattice relaxation of paramagnetic ions was observed in these experiments, as well as in the studies of Zverev [74] and of the present author. Analysis of the temperature and concentration dependences of the experimental data on the rate of relaxation indicated, as in the case of $K_3(Fe,Co)(CN)_6$, the important role of exchange-coupled ions in the transfer of the excitation energy from the spins to the lattice.

CHAPTER IV

Apparatus for the Observation of Electron Paramagnetic Resonance at 42 MHz and Below 1°K

Recently, many investigations of electron paramagnetic resonance and of paramagnetic relaxation have been carried out at low temperatures. The use of liquid helium has made it possible to cool samples to 1-1.5°K. However, in the investigation of certain problems, for example, the spin-lattice relaxation mechanism in paramagnetic crystals or the mechanism of magnetic resonance in substances with weak exchange interactions (Curie temperature below 1°K), it is interesting to carry out experiments on electron spin resonance below 1°K.

Only a few experiments have been carried out so far on the electron spin resonance at these extremely low temperatures. Cooling has been achieved by the adiabatic demagnetization method. The first experiment was carried out at 9600 MHz on the paramagnetic relaxation of Nd^{3+} and Ce^{3+} ions in LaMg nitrate [37]. Two other experiments are described in the present paper. One of them [32] is concerned with the observation of the paramagnetic-antiferromagnetic phase transition in a molecular crystal of the stable organic free radical α,α-diphenyl-β-picrylhydrazyl, whose Curie temperature is $\Theta \sim 0.35°K$. The other involves measurements of the paramagnetic relaxation [65, 66] in $K_3(Fe,Co)(CN)_6$. These experiments were carried out by the present author and his colleagues in the meter band of electromagnetic wavelengths. We shall now describe the apparatus used in these experiments [59].

§1. General Layout of the Apparatus

The general layout of a spectrometer used to observe electron paramagnetic resonance at very low temperatures in the meter band of wavelengths ($\lambda \approx 6$ m) is shown in Fig. 13.

The main unit of the apparatus is a glass helium cryostat with a liquid-nitrogen jacket. An LC oscillatory circuit is placed in the cryostat. The coil of this circuit is placed in a vacuum chamber immersed in liquid helium; the coil encloses the crystal or crystalline powder under investigation, cooled to $T < 1°K$. The coil is in good thermal contact with the liquid helium. A coaxial transmission line, which connects the ends of the coil to the sealed cap of the cryostat, is used as the capacitance in the circuit. The sample is connected by a copper heat conductor to a paramagnetic salt pellet cooled by the adiabatic demagnetization method. The LC circuit is the grid circuit of an autodyne oscillator, working at a fixed frequency of 42 MHz. Resonance absorption of the rf magnetic field energy by the sample (this energy is provided by the coil) reduces the Q factor of the circuit and is recorded as a change in the amplitude of the oscillator. The detected signal is amplified by a dc amplifier and observed on the screen of an oscilloscope. The amplitude of the oscillator signal is measured with a voltmeter. A calibrator (reference) circuit, described in [36], is used to check the linearity of the spectrometer and to

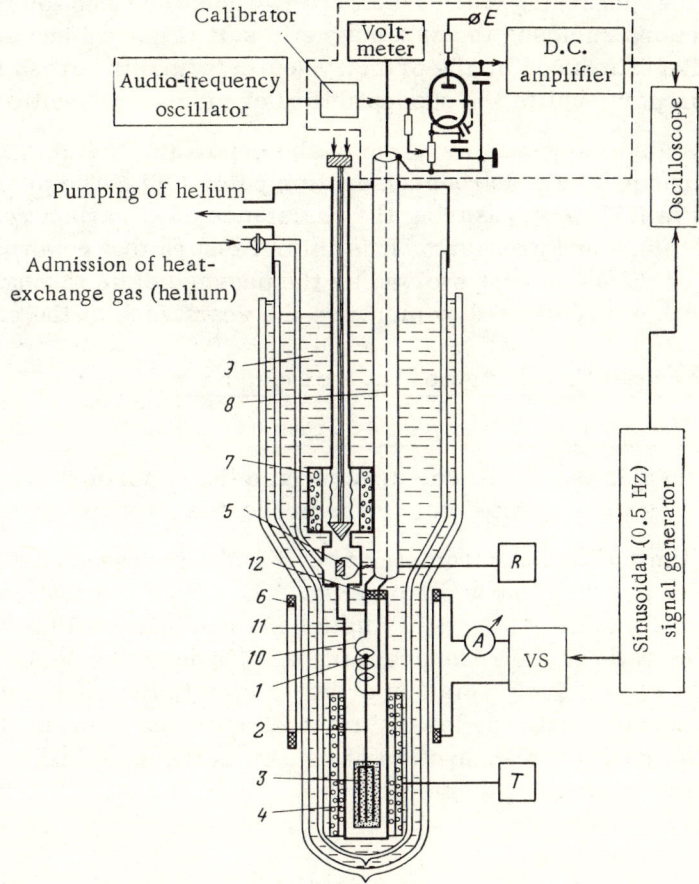

Fig. 13. EPR spectrometer for measurements at very low temperatures (<1°K): 1) investigated sample; 2) copper heat conductor; 3) paramagnetic salt used to cool the sample; 4) coil used to determine the temperature by measuring the susceptibility of the paramagnetic salt; 5) carbon resistor for measuring the vacuum in the system; 6) Helmholtz coils; 7) carbon adsorption pump; 8) coaxial transmission line in the grid circuit of autodyne oscillator; 9) glass helium cryostat with liquid-nitrogen jacket; 10) LC coil in the oscillator circuit; 11) vacuum chamber; 12) vacuum seal of central core of the coaxial transmission line.

measure the amplitude of the absorption signal in relative units. The splitting of the energy levels in the sample, corresponding to the frequency of the hf magnetic field, is induced by a static magnetic field generated by Helmholtz coils. These Helmholtz coils are supplied from a stabilized voltage source (VS). The value of the static magnetic field is found from the current through the coils (measured with an ammeter A). The static magnetic field is modulated sinusoidally at a frequency of 0.5 Hz in order to make it possible to observe the resonance absorption signal on the oscilloscope screen. This modulation of the static field is achieved by feeding a sinusoidal signal of 0.5 Hz frequency to the stabilizer circuit of the voltage supply for the Helmholtz coils. The same sinusoidal signal is used for the synchronization of the oscilloscope display.

The temperature of the sample under study is assumed to be equal to the temperature of the paramagnetic salt and its value is found by measuring the static magnetic susceptibility of the salt by a measuring coil and a bridge circuit (T), which draws power from the line supply.

The inner winding of the measuring coil consists of two identical sections connected in opposition; one of these sections encloses the paramagnetic salt (Figs. 13 and 15). The bridge circuit measures the voltage induced by the primary (outer) winding across the secondary winding of the coil, which is proportional to the susceptibility of the paramagnetic salt.

Figure 13 shows also the vacuum system of the cryostat. Helium gas, used as a heat-exchange medium, is pumped by a carbon adsorption pump [76]. The pressure of the helium gas is determined, as in [77], by measuring the resistance of a carbon resistor, which depends strongly on the temperature and pressure. It is this pressure that governs the rate of transfer (to the liquid helium) of the Joule heat evolved by the passage of dc through the resistor (R in Fig. 13 denotes the bridge circuit used to measure the resistance of the carbon resistor).

§ 2. Special Features of Magnetic Resonance Experiments at Very Low Temperatures

We shall now consider the conditions under which the experiments on magnetic resonance at very low temperatures were carried out by the present author and Prokhorov [32].

The sensitivity limit of the spectrometer at T = 4.2°K corresponded to a minimum number of $2 \cdot 10^{17}$ electron spins per unit line width of 1 Oe. The signal from a molecular crystal of DPPH of 25 mg weight, containing $N \approx 3 \cdot 10^{19}$ spins and exhibiting a line $\Delta H_0 \approx 5$ Oe wide, was observed at 4.2°K. The ratio of the resonance line amplitude to the noise amplitude was about 30. The sample had to be relatively small in order to satisfy the inequality $\Delta r/r \ll 1$ (here r is the resistance which represents the losses in the circuit; Δr represents the additional losses in the circuit due to the resonance absorption of energy in the sample), which had to be obeyed throughout the investigated range of temperatures in order to ensure the linear operation of the spectrometer:

$$\frac{\Delta r}{r} = 4\pi \xi \chi'' Q. \tag{4.1}$$

Let us calculate this ratio for a sample of 25 mg weight at T = 0.5°K and χ'' close to its maximum value [32]. The measured Q factor of the LC circuit is $Q \approx 300$. The imaginary component of the magnetic susceptibility in the absence of saturation is $\chi'' = \chi_0(H_0/\Delta H)$, where the ratio of the resonance value of the field to the line width is $H_0/\Delta H_0 \approx 2.5$ [32] and the static magnetic susceptibility $\chi_0 = Ng^2\beta^2/4k(T + \Theta)$ is used for spin $s = \frac{1}{2}$. Substituting the values $N = 3 \cdot 10^{19}$ spins, $g \approx 2$ for DPPH, $\Theta = 0.35°K$ [32], and the space factor $\xi = 1/V$ (where $V = 3.2$ cm^3 is the volume of the coil), we find that $\Delta r/r \approx 0.06$. This estimate is in agreement with the experimentally observed small change in the amplitude of the oscillator signal due to the resonance absorption of energy in the sample.

Chemically pure iron-ammonium alum $FeNH_4(SO_4)_2 \cdot 12H_2O$ was used as the paramagnetic salt in the adiabatic demagnetization process. The temperature to which this alum could be cooled depended on the initial temperature T_0, from which the magnetic cooling was started, and on the magnetic field H, used in the isothermal magnetization. This temperature was given by the ratio $T/T_0 \approx 460/H$ [78]. Under our conditions ($T_0 = 1.65°K$, H = 7.5 kOe), the calculated value of T was 0.1°K. The experimentally obtained value was also T = 0.1°K. When a more efficient pump was employed for the removal of helium vapor, we were able to reduce the temperature of liquid helium to 1.35°K and to cool the paramagnetic salt to $T \approx 0.08°K$.

A crystal of DPPH was cooled using an alum pellet, sufficiently large to ensure that the heat capacity of the alum was considerably higher than the heat capacity of the cooled crystal, and that reliable thermal contact was established between the cooled sample and the alum. The alum pellet was cylindrical, 6 cm long and 1.5 cm in diameter. It was compressed until its density reached 1.7 g/cm^3. The heat capacity of such a pellet was C $\approx$ 1.7·10^6 erg/deg K at T = 0.15°K [78] and the heat capacity of the lattice of a DPPH crystal of 25 mg weight (Debye temperature 50°K [24]) was $C_L \approx$ 25 erg/deg K at T = 1.5°K. Thus, throughout the investigated temperature range, the heat capacity of the alum pellet was several orders of magnitude higher than the heat capacity of the spin system of the DPPH sample, which was primarily due to exchange interactions. The heat capacity of the spin system was estimated from the formula

$$C_{\text{exch}} = 2R\left(\frac{J}{kT}\right)\exp\left(-\frac{J}{kT}\right)\left[1 + \exp\left(-\frac{J}{kT}\right)\right]^{-2}.$$

At T $\approx$ 0.4°K, at which J/kT $\approx$ 1 [32], C_{exch} had a maximum equal to $C_{\text{exch}} \approx$ 1.7·10^3 erg/deg K for a crystal of 25 mg weight. These estimates indicated that the size of the alum pellet imposed practically no restriction on the weight of the cooled crystal and, therefore, on the amplitude of the resonance signal. Even when the weight of the sample was increased to a value corresponding to $\Delta r/r \approx$ 0.5-1, i.e., corresponding to a change in the amplitude of the oscillator signal comparable with its original amplitude (our resonance signal was known to be smaller than such a change) the necessary relationship between the heat capacities was not violated.

Good thermal contact was maintained throughout the investigated range of temperatures, as supported by the following two observations. First, we found that the time for the establishment of thermal equilibrium between a sample and the alum at T $\approx$ 0.2°K was short (about 3 sec). This time was deduced from observations of the cooling of a sample, which was heated first by the alternating field (20 Oe, 50 Hz) of the Helmholtz coils by ΔT $\lesssim$ 0.1°K above the temperature of the alum pellet. The initial rise of the temperature of the sample and its subsequent fall were deduced from the strong temperature dependence of the amplitude of the resonance signal of DPPH, observed at T $\approx$ 0.2-0.4°K (the resonance line disappeared below 0.2°K) [32]. Secondly, throughout the 0.4-4°K range, the area under the resonance absorption line obeyed the law const/(T + Θ); the temperature T below 1.65°K was deduced from the temperature of the alum.

The determination of the temperature dependences of the characteristics of the EPR spectrum in the range 0.15-1.5°K was carried out during the heating of the sample. The rate of heating was slow. The temperature of the paramagnetic salt pellet increased for the following reasons: 1) transfer of heat from liquid helium by conduction through the suspension of the paramagnetic salt; 2) evolution — in the copper heat conductor — of the Joule heat of the Foucault currents induced by alternating magnetic fields; 3) absorption of heat from the sample, which was heated by the resonance absorption of the hf magnetic field energy from the oscillator circuit. During these experiments the following alternating fields were operating continuously: a) the static magnetic field, sinusoidally modulated at a frequency of 0.5 Hz and an amplitude of 26 Oe; b) the alternating field of the measuring coil (50 Hz) whose amplitude was 0.3 Oe at T $\approx$ 0.15-0.25°K and 3 Oe at T > 0.25°K; c) the alternating field of the oscillator coil whose frequency was 5·10^7 Hz and whose amplitude did not exceed $H_1 \approx$ 0.02 Oe. The amplitude of the oscillator-coil field was found from the relationship $CU_0^2 = H_1^2 V/4\pi$ and it corresponded to a voltage of $U_0 \approx$ 1 V across the circuit, a capacitance C $\approx$ 10 pF, and a coil volume V $\approx$ 3.2 cm^3. The variation of the temperature with time in two experiments is shown in Fig. 14.

It is evident from Fig. 14 that: the rate of rise of temperature at the beginning of each experiment was 5·10^{-4} deg K/min; after an hour it rose to 2.5·10^{-3} deg K/min, and after two hours it increased to 2.5·10^{-2} deg K/min. The temperature rose from 0.10 to 0.20°K in 1 hr; after 2 hr it reached 0.9°K, and after 2.5 hr, it rose to T $\approx$ 1.5°K. These conditions were satisfactory for magnetic resonance experiments. The slow modulation of the static magnetic field made a small contribution to the heating, represented by a heating rate of 10^{-5} deg K/min at T $\approx$ 0.2°K, while the actually observed rate of heating at this rate was 0.1 deg K/min for a field of 50 Hz frequency and 20 Oe amplitude (this field was used for deliberate heating). The power dissipated as Joule heat was proportional to the square of the frequency of the field (the depth

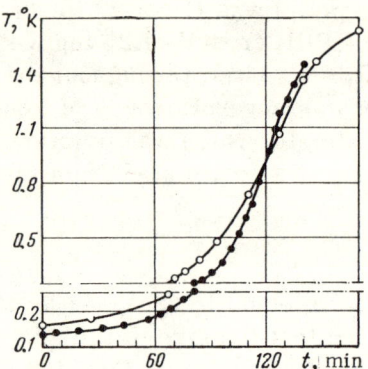

Fig. 14. Dependence of the temperature of the investigated sample on time.

of penetration of the magnetic field was $\delta \approx 2$ cm at $\nu = 0.5$ Hz and $\delta \approx 0.2$ cm at $\nu = 50$ Hz for a copper heat conductor which was much longer — along the direction of the field — than the heat conductor actually used). The resonance absorption in the sample and the Joule heat of the Foucault currents generated by the hf field of the oscillator were not important for the oscillator signal amplitudes from 0.1 to 1.0 V, because changes in the oscillator amplitude within the limits employed did not alter greatly the rate of heating. The reported data on the rate of rise of temperature in the system were obtained in our first experiments. In subsequent experiments, the rate of heating was considerably lower and the duration of the period at which the apparatus remained at a temperature $T < 1°K$ (3-4 hr) was governed by the amount of liquid helium in the vessel and not by the rate of heating of the system. The paramagnetic salt had to be heated specially in order to record the temperature dependences.

§3. Construction of the Principal Units of the Apparatus

Let us now consider some parts of the apparatus in detail. Figure 15 shows the construction of the vacuum chamber in which the crystal to be cooled was placed. The suspension of the paramagnetic salt pellet by vertical thin-walled stainless-steel tubes, attached to the cover of the chamber, was the same as that described in [77]. This construction is simple and very rigid and ensures that the absorption of heat from the outside is small. The problem of obtaining good thermal contact between an organic molecular crystal and the copper heat conductor was solved as follows: the crystal was surrounded by a copper-wire "brush" and glycerin was poured into the crystal container. The glycerin froze and provided a good thermal contact. The brush consisted of 100 copper wires, 0.1 mm in diameter and 20 mm long, which were silver-soldered to the upper part of the heat conductor. The total contact area between the copper and the frozen glycerin was about 6 cm^2. The surface area of the copper plates embedded in the alum pellet was almost 60 cm^2; these plates were silver-soldered to the lower part of the heat conductor. The upper and lower parts of the heat conductor were joined with a low-melting-point solder. A copper brush was also used in [79] for the magnetic cooling of alcohol solutions of paramagnetic salts. In our experiments, the central core of the coaxial transmission line entered the vacuum chamber through a platinum tube, 0.5 mm in diameter, which was sealed to the glass wall of the chamber by a copper–glass–platinum joint.

§4. Measurement of the Spin-Lattice Relaxation Rate at $T < 1°K$

We shall now consider the measurement of the spin-lattice relaxation time at very low temperatures. We shall assume that the precision of the measurement of τ is satisfactory when the resonance signal amplitude decreases to half its value, corresponding to the following saturation factor (spin $s = \frac{1}{2}$):

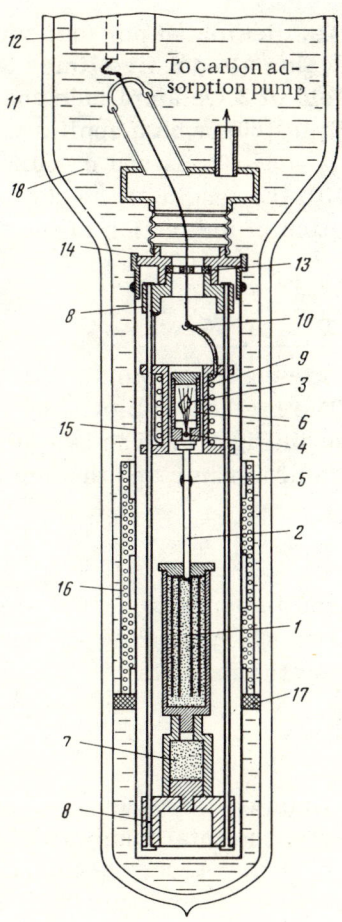

Fig. 15. Construction of the sample container: 1) compacted iron ammonium alum; 2) lower part of copper heat conductor; 3) cooled crystal; 4) upper part of copper heat conductor; 5) joint; 6) glycerin; 7) alum pellet used for shielding; 8) suspension system for alum pellet, heat conductor, and sample; 9) copper-wire coil on Plaxiglas former; 10) joint between coil end and platinum core of coaxial line; 11) glass-vacuum seal for central core of coaxial line; 12) coaxial line; 13) Teflon centering-disk; 14) seal made of low-melting-point solder; 15) stainless-steel vacuum chamber; 16) measuring coil, shown in its position relative to alum pellet; 17) plastic-foam centering ring; 18) liquid helium.

$$X = \left(1 + \frac{\gamma H_1^2 \tau_1}{4 \Delta H_0}\right)^{-1} = 0.5, \tag{4.2}$$

where γ is the gyromagnetic ratio. An increase in H_1^2 increases the absorption in the sample but it is limited by the maximum permissible heating rate. The heating of a paramagnetic salt pellet by the resonantly absorbed power P proceeds at the rate

$$\frac{dT}{dt} = \frac{P\alpha}{C_1}. \tag{4.3}$$

The heat capacity C_1 of an alum pellet 18 g in weight is $C_1 = 4 \cdot 10^4/T^2$ [78]. The power is given by $P = 0.5 \omega_0 H_1^2 \chi'' = \pi \nu_0 H_1^2 \left(\frac{H_0}{\Delta H_0}\right) X \left(\frac{Ng^2\beta^2}{4kT}\right)$. This equation is valid for spin s = ½. The coefficient α is equal to the ratio of the saturation time of the sample being investigated and the duration of the observation period. In the pulse saturation case, this coefficient is equal to the ratio of the duration of the saturating pulse and the pulse repetition period; in the continuous saturation case, it is equal to the ratio of twice the time of scan of the line and the modulation period. Let us assume that $\alpha \approx 0.1$. For a DPPH crystal of 25 mg weight, we find that dT/dT = $4.3 \cdot 10^{-3} H_1^2$ under the following conditions: $\nu_0 = 5 \cdot 10^7$ Hz, T = 0.5°K, ΔH_0 = 7.5 Oe [32]. If the maximum permissible heating rate (taken from Fig. 14) is 0.016 deg K/min, we find that $H_1 \approx 0.25$ Oe. According to Eq. (4.2), this field can be used to measure τ of the DPPH sample considered if this time constant is not shorter than about 10^{-5} sec. According to the relationship $CU_0^2 = H_1^2 V/4\pi$, this corresponds to an oscillation amplitude $U_0 \approx 12$ V. The operation of the oscillator circuit at this level would increase the amount of Joule heat Q_0, dissipated per unit time by the Foucault currents induced in the copper heat conductor, to a value almost five times higher than the rate of absorption of heat from the measuring coil field of 3 Oe amplitude and 50 Hz frequency. This estimate is based on the relationship $Q_0 \propto H_1^2 \sqrt{\nu}$ [80], which applies when the penetration depth of the alternating fields ($\delta \approx 2 \cdot 10^{-4}$ cm at $\nu_0 = 5 \cdot 10^7$ Hz and $\delta \approx 0.2$ cm at $\nu = 50$ Hz) in the copper heat conductor is small compared with the length of the conductor. We can reduce Q_0 by decreasing the value of the ratio of the duration of saturation of the sample to the observation period.

§ 5. Method for Measuring Temperatures at $T < 1°K$

The temperature of a paramagnetic salt pellet in magnetic cooling experiments is usually deduced from the static magnetic susceptibility of the salt. The method employed by us was the same as that used in [77]. The basic electric circuit is shown in Fig. 16. The two (primary and secondary) windings of the measuring coil are wound on the same former and the primary winding is fed with an alternating current

$$i = I_0 \sin \omega t.$$

The secondary winding consists of two identical sections, wound in opposition. One section encloses the paramagnetic salt pellet. The voltage across the secondary winding V_2, equal to the sum of the voltages across the two sections $V_2' + V_2''$, is given by the formula

$$V_2 = 0.4\pi \left(\frac{n_1}{l_1}\right) S I_0 \omega \left[(n_2' - n_2'') + n_2' 4\pi\chi_0\right] \cos \omega t,$$

where $n_1/l_1 = 50$ is the number of turns per centimeter in the primary winding; $S = 7$ cm^2 is the cross sectional area of the coil; n_2' and n_2'' are, respectively, the total numbers of turns in the first and second sections of the secondary winding; $n_2' \approx n_2'' = n_2 = 1500$; $\chi_0 = 0.014/T$ [78] is the static magnetic susceptibility of the paramagnetic salt pellet. The magnetic field generated by the primary winding is assumed to be uniform throughout the length of the coil. A uniform field is obtained by increasing the number of turns at the ends of the coil. If the sections of the secondary winding have the same numbers of turns ($n_2' = n_2'' = n_2$), the amplitude of the measured voltage is

$$V_2 = 0.4\pi \left(\frac{n_1}{l_1}\right) I_0 S \omega n_2 4\pi\chi_0. \tag{4.4}$$

When the frequency is $\omega = 2\pi\nu = 314$ Hz (ac power supply) and the amplitude of the current in the primary winding is $3\sqrt{2}$ mA, the calculated value of V_2 at $T = 1°K$ is

$$V_2 = 2 \cdot 10^4 \text{ emu} = 0.0002 \text{ v}.$$

This is in reasonable agreement with the experimental value of 0.0020 V when the effective current in the primary coil is 30 mA. The voltage V_2 is measured with an ac potentiometer of the R-56 type. The voltage V_2 is shifted in phase relative to the reference voltages X and Y of the potentiometer. Therefore,

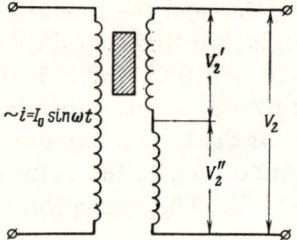

Fig. 16. Circuit for measuring static ac magnetic susceptibility. The shaded rectangle represents a paramagnetic salt pellet.

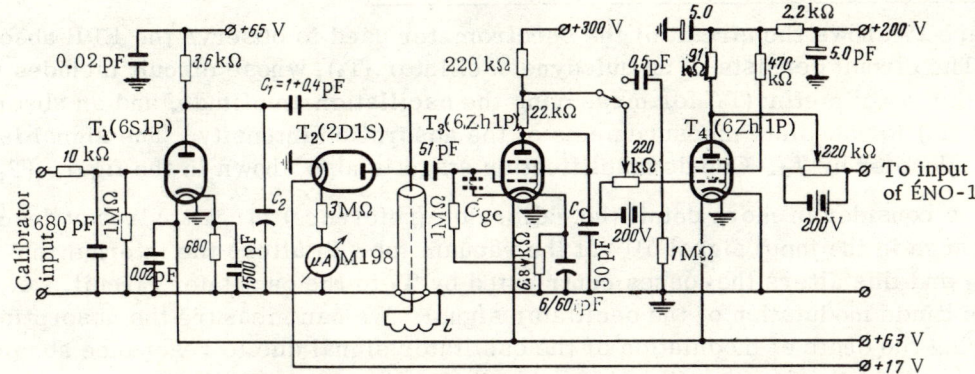

Fig. 17. Electric circuit of the spectrometer used to observe magnetic resonance absorption at 42 MHz.

$$V_2 = \sqrt{(X - X_0)^2 + (Y - Y_0)^2},$$

where X_0 and Y_0 are potentials which appear because the two sections of the secondary winding are not identical.

The magnetic temperature T^* can be deduced from the measured susceptibility using the relationship

$$\chi = \frac{C}{T^*}.$$

It follows from Eq. (4.4) that

$$T^* = \frac{\text{const}}{V_2}. \tag{4.5}$$

The constant in Eq. (4.5) is determined experimentally from independent measurements of T^* and V_2 in the liquid helium temperature range 4.2–1.35°K.

Next, the magnetic temperature T^* is extrapolated to obtain the temperature of a spherical sample $T^{(*)}$ [78]:

$$T^{(*)} = T^* + \Delta, \tag{4.6}$$

where

$$\Delta = \left(\frac{4\pi}{3} - \varepsilon\right)\frac{C}{V}.$$

Here ε is the demagnetization factor of the paramagnetic salt pellet; C/V is the susceptibility of 1 cm³ of the salt at $T = 1$°K (the density of the compacted pellet is equal to the normal density of a crystal). It is assumed that the demagnetization factor of a cylindrical salt pellet, 6 cm long and 1.5 cm in diameter, is $\varepsilon = 1$, i.e, exactly the same as for an ellipsoid of revolution with the length-to-diameter ratio $l/d = 4$ [78]. Consequently, in our experiments $\Delta = 0.0494$°K. Finally, the temperature of a spherical sample $T^{(*)}$ is transformed to the thermodynamic temperature T using the appropriate graph given in [81]. It must be pointed out that the difference between $T^{(*)}$ and T is important only at $T < 0.3$°K.

§6. Electric Circuit of the Spectrometer

Figure 17 shows the circuit of the spectrometer used to observe the EPR absorption at 42 MHz. The circuit consists of an autodyne oscillator (T_3), whose circuit includes the investigated sample, a voltmeter (T_2) for measuring the oscillation amplitude, and an absorption signal calibrator (T_1) for absolute measurements of the absorption intensity. The signal is detected by the tube denoted by T_3. One dc amplification stage is also shown in the figure (T_4).

Let us consider in more detail the calibrator (reference) circuit, already described in [36]. A change in the input signal δU_g at the vacuum tube T_1 alters the internal impedance of the tube T_1 and this alters the losses contributed by T_1 to the oscillator circuit, i.e., it gives rise to amplitude modulation of the oscillator signal. We can measure the absorption intensity by comparing the depth of modulation of the oscillator signal due to resonance absorption in the sample with the depth of modulation induced by the calibrator circuit.

The admittance, introduced by the calibrator in the LC circuit, is

$$\Delta G = \left(\frac{C_1}{C_2}\right)^2 \frac{\Delta R_i}{R_i^2} = \left(\frac{C_1}{C_2}\right)^2 \frac{1}{R_i^2} \frac{dR_i}{dU_g} \Delta U_g. \qquad (4.7)$$

In the derivation of Eq. (4.7) we have assumed that $C_1 \ll C_2$, $C_1 \ll C$, $\omega R_a' C_2 \gg 1$, where $R_a' = R_i R_a / (R_i + R_a)$ and ΔU_g is the voltage between the grid and the cathode.

Standard relationships for the input control signal δU_g and the anode current i_a

$$\delta U_g = \Delta U_g + i_a R_c,$$

$$i_a = \frac{\mu \Delta U_g}{R_c + R_a + R_i}$$

can be used to show easily that

$$\Delta U_g = \delta U_g \frac{1 + \dfrac{R_c + R_a}{R_i}}{1 + R_c S + \dfrac{R_c + R_a}{R_i}},$$

i.e.,

$$\Delta G = \text{const} \, (\delta U_g). \qquad (4.8)$$

This relationship is basic. It shows that the intensity of the resonance absorption signal can be calibrated using the amplitude of the signal at the calibrator circuit input.

We have used the following notation in this section: R_i is the internal resistance of the tube T_1; S is the slope of the anode–grid characteristic of the tube; R_a and R_c are, respectively, the anode and cathode loads; $\mu = SR_i$; C_1 and C_2 are the capacitances shown in the circuit of Fig. 17.

The calibrator circuit just described can be used throughout the full range of changes in the absorption signal amplitude encountered in our investigation.

Conclusions

The most important results obtained in our investigation are as follows.

1. The main result of the experiments on the free radical α,α-diphenyl-β-picrylhydrazyl is the disappearance (below 1°K) of the paramagnetic resonance absorption line in all the investigated samples. This is clear evidence of a change in the energy spectrum corresponding to a transition from paramagnetism to an ordered magnetic state. This conclusion is supported also by the good agreement of the information deduced from measurements of the static magnetic susceptibility of a fine-grained powder of the radical, its line width, values of the isotropic exchange integral J, and the transition temperature. The rapid broadening, associated with the appearance of magnetic order at T < 1°K, also indicates a phase transition. The temperature interval in which this transition is observed is in good agreement with the interval in which the integral intensity of the EPR line decreases rapidly. This is the first time that a transition from paramagnetism to antiferromagnetism was observed in molecular crystals of organic free radicals. The data on the line width and the Curie temperature reported by other workers are used to draw some conclusions on the structure of the spin system of DPPH in the paramagnetic region. Further experimental work is required to establish the cause of the difference between the experimental results obtained for fine-grained powder and for large single crystals.

2. The mechanism of the participation of exchange-coupled ions (ion pairs) in the spin-lattice interaction of isolated paramagnetic ions in magnetically dilute systems was investigated using paramagnetic crystals of $K_3(Fe,Co)(CN)_6$.

Measurements of the spin-lattice relaxation rate of Fe^{3+} ions in the lattice of $K_3Co(CN)_6$ at 42 MHz, carried out in the temperature range T = 4.2-0.1°K on samples with iron ion concentrations $f = 10^{-3}$, indicate a strong concentration dependence of the relaxation rate and a deviation from the Van Vleck–de Krönig temperature dependence. The energy spectrum of ion pairs known from other investigations and a calculation of the spin-phonon interaction of ion pairs show that, although such pairs are unimportant in strong magnetic fields at high frequencies, they play the dominant role in the spin-lattice relaxation mechanism at low frequencies. Estimates are given of the order of magnitude of the probabilities of the relaxation transitions of Fe^{3+} ions in $K_3Co(CN)_6$. These estimates are in agreement with the experimental data. An approximate but quantitative comparison is made (for the first time) between the experimental data and the theory of relaxation involving ion pairs. This comparison shows that such a mechanism is important. A detailed theoretical analysis of the experimental data is possible because of the simplicity of the investigated material, $K_3(Fe,Co)(CN)_6$.

3. The results of other workers [4, 5, 47] are used to discuss how a generalization of the Van Vleck–de Krönig theory of the paramagnetic spin lattice relaxation [2, 3] can be used to explain the experimentally observed anomalies in the one-phonon spin-lattice relaxation. This generalization is based on the inclusion of the spin-spin degrees of freedom in the spin-lattice relaxation mechanism [4]. The discussion is applied specifically to strong (compared with the Zeeman interactions) spin-spin exchange interactions between paramagnetic particles, investigated in the present study. It may be concluded that the theory of paramagnetic relaxation, at least for the simplest systems (i.e., systems whose spin Hamiltonian includes only the Zeeman, exchange, and magnetic-dipole energies of spins), is now complete.

4. The experimental part of the present paper reports new methods. Until now, the extension of measurements of EPR in the direction of low temperatures has been limited by the temperature of liquid helium (T > 1°K). Our experiments, like the study of the spin-lattice relaxation in rare-earth salts reported in [37], are the first attempts to investigate the EPR at extremely low temperatures T < 1°K.

The investigations reported in the present paper were carried out in the L. I. Mandel'shtam and N.D. Papaleksi Oscillations Laboratory of the P.N. Lebedev Physics Institute of the Academy of Sciences of the USSR. The project lasted from March 1960 until January 1964. Magnetic cooling to T = 0.1°K was achieved in November 1961. The first successful observations of the EPR spectrum of DPPH free radicals at T < 1°K were carried out in April 1962. The experiments on the spin-lattice relaxation in $K_3(Fe,Co)(CN)_6$ were performed between October 1962 and June 1963.

The author is profoundly grateful to A. M. Prokhorov for his great help over many years, as well as for his always lively and constructive comments on the results of the present work. The author is deeply indebted to B. M. Samoilov for access to the detailed information on the magnetic cooling technique accumulated in his laboratory. Many constructions, measuring circuits, and technical procedures used in the present experiments were developed in Samoilov's laboratory. The author offers his sincerest thanks to B. V. Ershov for his initiative, inventiveness, and effort in helping to prepare the apparatus, in carrying out the measurements, and in the analysis of the experimental data. The author is also indebted to K. K. Svidzinskii for his timely correction of an error in the theoretical calculations of the probabilities of spin-lattice transitions in potassium cobalticyanide. E. N. Bol'shakov, D. K. Bardin, and D. A. Klepikov kindly helped the author in the construction and preparation of the apparatus, P. A. Tseitler prepared DPPH in the form of a fine powder, and V. S. Borodachev helped in the preparation of the manuscript for press.

References

1. L. S. Singer and C. Kikuchi, J. Chem. Phys., 23:1738 (1955).
2. R. de Krönig, Physica, 6:33 (1939).
3. J. H. Van Vleck, Phys. Rev., 57:426, 1052 (1940).
4. N. Bloembergen and S. Wang, Phys. Rev., 93:72 (1954).
5. J. H. Van Vleck, Quantum Electronics (ed. by C. T. Townes), Columbia Univ. Press, New York (1960), p. 392.
6. N. Bloembergen and P. S. Pershan, Advances in Quantum Electronics (Proc. Second Intern. Conf., Berkeley, 1961), (ed. by J. R. Singer), Columbia Univ. Press, New York (1961), p. 373.
7. J. H. Van Vleck, Advances in Quantum Electronics (Proc. Second Intern. Conf., Berkeley, 1961), (ed. by J. R. Singer), Columbia Univ. Press, New York (1961), p. 388.
8. J. C. Gill and R. J. Elliott, Advances in Quantum Electronics (Proc. Second Intern. Conf., Berkeley, 1961), (ed. by J. R. Singer), Columbia Univ. Press, New York (1961), p. 399.
9. j. C. Gill, Proc. Phys. Soc. (London), 79:58 (1961).
10. H. Statz, L. Rimai, M. J. Weber, G. A. DeMars, and G. Koster, J. Appl. Phys., 32(Suppl.): 218S (1961); M. J. Weber, L. Rimai, H. Statz, and G. DeMars, Bull. Am. Phys. Soc., 6:141 (1961).
11. S. A. Al'tshuler, Zh. Éksp. Teor. Fiz., 43:2318 (1962).
12. I. Waller, Z. Physik, 79:370 (1932).
13. B. M. Kozyrev and S. G. Salikhov, Dokl. Akad. Nauk SSSR, 58:1023 (1947).
14. A. N. Holden, C. Kittel, F. R. Merritt, and W. A. Yager, Phys. Rev., 77:147 (1950).
15. C. H. Townes and J. Turkevich, Phys. Rev., 77:148 (1950).
16. D. J. E. Ingram, Free Radicals as Studied by Electron Spin Resonance, Butterworths, London (1958).

17. S. A. Al'tshuler and B. M. Kozyrev, Electron Paramagnetic Resonance [in Russian], Fizmatgiz, Moscow (1961).
18. H. Katz, Z. Physik, 87:238 (1933).
19. E. Müller, I. Müller-Rodloff, and W. Bunge, Annalen der Chemie, 520:235 (1935).
20. J. Turkevich and P. W. Selwood, J. Am. Chem. Soc., 63:1077 (1941).
21. L. S. Singer and E. G. Spencer, J. Chem. Phys., 21:939 (1953).
22. H. J. Gerritsen, R. Okkes, H. M. Gijsman, and J. van den Handel, Physica, 20:13 (1954).
23. J. P. Goldsborough, M. Mandel, and G. E. Pake, Phys. Rev. Letters, 4:13 (1960).
24. J. P. Goldsborough, M. Mandel, and G. E. Pake, Proc. Seventh Intern. Conf. on Low Temperature Physics, University of Toronto, Canada, 1960 (ed. by G. M. Graham and A. C. Hollis Hallett), Univ. of Toronto Press (1961), p. 702.
25. R. S. Rhodes, J. H. Burgess, and A. S. Edelstein, Phys. Rev. Letters, 6:462 (1961).
26. A. S. Edelstein and M. Mandel, J. Chem. Phys., 35:1130 (1961).
27. M. E. Anderson, R. S. Rhodes, and G. E. Pake, J. Chem. Phys., 35:1527 (1961).
28. J. H. Burgess, R. S. Rhodes, M. Mandel, and A. S. Edelstein, J. Appl. Phys., 33(Suppl.):1352 (1962).
29. W. Duffy Jr., J. Chem. Phys., 36:490 (1962)
30. H. Sato, A. Arrott, and R. Kikuchi, J. Phys. Chem. Solids, 10:19 (1959).
31. F. M. Johnson and A. H. Nethercot Jr., Phys. Rev., 114:705 (1959).
32. A. M. Prokhorov and V. B. Fedorov, Zh. Éksp. Teor. Fiz., 43:2105 (1962).
33. A. M. Prokhorov and V. B. Fedorov, Zh. Éksp. Teor. Fiz., 44:1125 (1963).
34. R. B. Griffiths, Phys. Rev., 124:1023 (1961).
35. A. Van Itterbeek and M. Labro, Physica, 30:157 (1964).
36. G. D. Watkins, Thesis, Cambridge, Massachusetts (1952).
37. R. H. Ruby, H. Benoit, and C. D. Jeffries, Phys. Rev., 127:51 (1962).
38. J. H. Van Vleck, Phys. Rev., 74:1168 (1948).
39. P. W. Anderson and P. R. Weiss, Rev. Mod. Phys., 25:269 (1953).
40. R. Kubo and K. Tomita, J. Phys. Soc. Japan, 9:888 (1954).
41. A. Abragam, The Principles of Nuclear Magnetism, Clarendon Press, Oxford (1961).
42. A. W. Hanson, Acta Cryst., 6:35 (1953).
43. A. I. Kitaigorodskii, Organic Chemical Crystallography, Consultants Bureau, New York (1961).
44. T. Nagamiya, K. Yosida, and R. Kubo, Adv. Physics, 4:1 (1955).
45. A. S. Borovik-Romanov, Progress in Science, Physicomathematical Sciences, Vol. 4, Antiferromagnetism [in Russian], Izd. AN SSSR, (1962).
46. Yu. S. Karimov and I. F. Shchegolev, Zh. Éksp. Teor. Fiz., 46:399 (1964).
47. N. Bloembergen, S. Shapiro, P. S. Pershan, and J. O. Artman, Phys. Rev., 114:445 (1959).
48. J. H. Van Vleck, Nuovo Cimento, Suppl., 6:1081 (1957).
49. R. D. Mattuck and M. W. P. Strandberg, Phys. Rev., 119:204 (1960).
50. R. Orbach, Proc. Roy. Soc. (London), A264:458 (1961).
51. M. I. Rodak, Fiz. Tverd. Tela, 6:521 (1964).
52. N. Bloembergen, E. M. Purcell, and R. V. Pound, Phys. Rev., 73:679 (1948).
53. A. K. Morocha, Fiz. Tverd. Tela, 4:2297 (1962).
54. B. W. Faughnan and M. W. P. Strandberg, J. Phys. Chem. Solids, 19:155 (1961).
55. M. Mandel, G. E. Pake, and J. P. Goldsborough, Bull. Am. Phys. Soc., 6:141 (1961).
56. É. M. Gasanov, A. M. Prokhorov, and V. B. Fedorov, Fiz. Tverd. Tela, 6:193 (1964).
57. J. P. Lloyd and G. E. Pake, Phys. Rev., 92:1576 (1953).
58. M. A. Garstens, L. S. Singer, and A. H. Ryan, Phys. Rev., 96:53 (1954).
59. V. B. Fedorov, Pribory i Tekh. Éksperim., 1963(4):98 (1963).
60. D. H. Paxman, Proc. Phys. Soc. (London), 78:180 (1961).
61. T. Bray, G. C. Brown Jr., and A. Kiel, Phys. Rev., 127:730 (1962).

62. I. V. Ovchinnikov, Fiz. Tverd. Tela, 4:2750 (1962).
63. P. P. Pashinin and A. M. Prokhorov, Zh. Éksp. Teor. Fiz., 40:49 (1961); Fiz. Tverd. Tela, 5:2722 (1963).
64. J. M. Baker, B. Bleaney, and K. D. Bowers, Proc. Phys. Soc. (London), B69:1205 (1956).
65. A. M. Prokhorov and V. B. Fedorov, Quantum Electronics (Proc. Third Intern. Conf., Paris, 1963), (ed. by P. Grivet and N. Bloembergen), Dunod, Paris (1964), and Columbia Univ. Press, New York (1964), p. 751.
66. A. M. Prokhorov and V. B. Fedorov, Zh. Éksp. Teor. Fiz., 46:1937 (1964).
67. F. R. McKim and W. P. Wolf, Proc. Phys. Soc. (London), B69:1231 (1956).
68. T. Ohtsuka, J. Phys. Soc. Japan, 15:939 (1960).
69. V. Barkhatov and G. Zhdanov, Structure Reports (for 1942-44), 9:209 (1955).
70. I. Dzyaloshinskii, J. Phys. Chem. Solids, 4:241 (1958).
71. T. Moriya, Phys. Rev., 120:91 (1960).
72. J. S. Griffith, The Theory of Transition-Metal Ions, Cambridge University Press (1961), p. 428.
73. L. D. Landau and E. M. Lifshitz, Quantum Mechanics, Non-Relativistic Theory, 2nd ed., Pergamon Press, Oxford (1965).
74. G. M. Zverev, Zh. Éksp. Teor. Fiz., 40:1667 (1961).
75. A. Rannestad and P. E. Wagner, Phys. Rev., 131:1953 (1963).
76. B. N. Samoilov, Dokl. Akad. Nauk SSSR, 86:281 (1952).
77. B. N. Samoilov, V. V. Sklyarevskii, and E. P. Stepanov, Zh. Éksp. Teor. Fiz., 38:359 (1960); B. N. Samoilov, V. V. Sklyarevskii, V. D. Gorobchenko, and E. P. Stepanov, Zh. Éksp. Teor. Fiz., 40:1871 (1961).
78. G. K. White, Experimental Techniques in Low-Temperature Physics, Clarendon Press, Oxford (1959).
79. A. R. Miedema, H. Postma, Miss N. J. van der Vlugt, and M. J. Steenland, Physica, 25:509 (1959).
80. L. D. Landau and E. M. Lifshitz, Electrodynamics of Continuous Media, Pergamon Press, Oxford (1960).
81. A. H. Cooke, Proc. Phys. Soc. (London), A62:269 (1949).